Roger Greeno

B.A. (Hons), M.C.I.O.B., F.I.O.P.

Construction processes
Level 1

Longman London and New York

Longman Group Limited
Longman House, Burnt Mill, Harlow
Essex CM20 2JE, England
Associated companies throughout the world

*Published in the United States of America
by Longman Inc., New York*

© Longman Group Limited 1983

First published 1983

British Library Cataloguing in Publication Data

Greeno, Roger
  Construction processes level 1.
  1. Building
  I. Title
  690   TH145

ISBN 0-582-41309-5

Printed in Singapore by
Huntsmen Offset Printing Pte Ltd.

General Editor – Construction and Civil Engineering

C. R. Bassett, B.Sc., F.C.I.O.B.

Formerly Principal Lecturer in the Department of Building and Surveying, Guildford County College of Technology

Books published in this sector of the series:

Building organisations and procedures   G. Forster
Construction site studies – production, administration and
    personnel   G. Forster
Practical construction science   B. J. Smith
Construction surveying   G. A. Scott
Materials and structures   R. Whitlow
Construction technology Volume 1   R. Chudley
Construction technology Volume 2   R. Chudley
Construction technology Volume 3   R. Chudley
Construction technology Volume 4   R. Chudley
Maintenance and adaptation of buildings   R. Chudley
Building services and equipment Volume 1   F. Hall
Building services and equipment Volume 2   F. Hall
Building services and equipment Volume 3   F. Hall
Measurement Level 2   M. Gardner
Measurement Level 3   M. Gardner
Structural analysis Level 4   G. B. Vine
Site surveying and levelling Level 2   H. Rawlinson
Economics for the construction industry   R. C. Shutt
Design procedures Level 4   J. M. Zunde
Design technology Level 5   J. M. Zunde
Architectural design procedures   C. M. H. Barritt
Construction technology for civil engineering
    technicians   P. L. Monckton
Environmental science   B. J. Smith, G. M. Phillips and M. Sweeney
Concrete technology Level 4   J. G. Gunning
Design of structural elements 1   A. G. Smyrell

Contents

Acknowledgements

We are grateful to the following for permission to reproduce copyright material:

British Standards Institution for extracts and diagrams based on material from past *British Standards and Codes of Practice* publications; Hepworth Iron Co for trade names and illustrations of 'Hepsleve', 'Hepline' and 'Hepseal'; the Controller of Her Majesty's Stationery Office for extracts from *Building Regulations* 1976 and illustrations from various *Building Research Establishment Digest*; ICI Fibres for trade name and illustration of 'Terram'; IMI Yorkshire Imperial Plastics Ltd, a subsidiary company of IMI Plc. for trade name and illustration of 'Marscar'; Key Terrain Ltd for trade name 'Terrain' and illustration of the 'Elostomeric Lip Seal' joint on UPVC underground pipes; National House-Building Council for extracts from its *Rules and Technical Requirements*; Plasclip Ltd for trade name and illustration of 'Plasclip' and some of their 'Reinforcement Spacers'; Radway Plastics Ltd for trade name and illustration of 'Dacatie'; Timber Research and Development Association for the 'TRADA' quality assurance mark on stress graded timber and illustrations from *TRADA Truss Design Manual*; Wavin Plastics Ltd for trade names 'Osma' and 'Roundline' and illustration of their rainwater system.

Author's acknowledgements

I would like to express my thanks to the series editor, Mr C. R. Bassett for providing the encouragement and opportunity to prepare this book. I am also extremely grateful to Mrs W. Whitney, the Building department librarian at Guildford College of Technology, for her assistance with provision of relevant information, and also John and Margaret Williams for their help with preparation of illustrations and typing the manuscript.

Chapter 1

Pre-construction processes

The decision to build normally precedes location and acquisition of suitable land for development. Site characteristics feature strongly in design and layout of buildings, but these should not dominate the design and restrict functional considerations. A site must be selected with properties which accommodate the proposed construction without undue expense and inconvenience.

Location of building land is normally through estate agents or private advertisements in the local papers and professional journals. Small plots are usually offered for sale at a fixed price or offers in the region of a certain price. Larger sites for estate and commercial development are generally auctioned. The period between advertising and completion of purchase extends for several weeks, providing ample time and opportunity for the purchaser to conduct a feasibility study. Several site visits and preliminary investigations will be undertaken to ascertain the effect of physical features and to explore the site's potential. The purchaser's solicitor or builder's legal representative will investigate the legislative aspects. These include rights of access, planning restrictions, preservation orders, nuisance factors, restrictive covenants, rights of light and easements.

Site investigation

Information accumulated from site visits should be tabulated and presented for easy reference in a report. The content should refer to the following factors:

2

(*a*) *The nearest town or city*. Proximity of schools, emergency services, entertainment, recreational, shopping, transport and employment facilities will considerably affect the economic viability of the site. The site value will be strongly influenced by these factors as will the potential selling price of the completed buildings.

(*b*) *Transport facilities between the site and neighbouring towns*. Local bus and train services will be important for the construction team and to the occupiers of the finished structure. The distance from the site to the nearest railway station should be noted.

(*c*) *The local authority or district council responsible for the site*. They provide information concerning site access, roads, development potential, location of building lines, planning application, conservation areas, preservation orders and location and tipping charges at spoil tips.

(*d*) *The level of the land, contours and topography*. This can be represented on a site plan by taking levels from a site datum point established from an ordnance bench mark, located on the Ordnance Survey map of the area (see Fig. 1.1). Trees, shrubs, existing buildings and other significant features should also be shown.

(*e*) *Trees and buildings*. These should be referred to the local authority to determine whether preservation orders exist or will be imposed. The restrictive effect on the layout of buildings can seriously affect development, and this should be represented in the land value.

(*f*) *Climatic and atmospheric conditions*. These are easily overlooked, but will have a significant effect on the style and material composition of the building. Air flow and snow are structural loadings affecting roof, wall and foundation construction. See Fig. 5.2 for an illustration of wind effect. Proposed tall or exposed structures could be affected by wind patterns, humidity, cloudiness or fog, and solar radiation could have an unpleasant effect if excessive, but is advantageous if exploited sensibly. Local industries may create atmospheric and noise pollution, and orientation, location and construction of buildings may require special consideration.

(*g*) *Demolition and alteration to existing buildings*. Consideration must be given to the methods and costs involved, resale value of second-hand materials and means of disposing of the surplus. Danger to neighbouring or adjacent buildings must also receive consideration with regard to means of protecting them. If shoring or underpinning is to be used as support to existing buildings, the owner's co-operation is essential, and liability for their condition accepted. Adequate insurance arrangements must be negotiated for this type of work and specialist sub-contractors consulted.

(*h*) *Access to the site*. It is important to ascertain whether tall or heavy plant can be transported along existing roads. The strength of bridges must be considered and if unusually shaped loads are anticipated the effect of low bridges and sharp corners must be foreseen. Access around the site often justifies temporary roads in the form of metal tracks. It is also useful to prepare the sub-base for permanent site roads for use as the basis for site transport.

(*i*) *The nature and composition of the subsoil*. Three factors to be determined are the natural level of water below the surface, the chemical content of the soil and the bearing capacity of the subsoil. Boreholes by hand or shell auger shown in Fig. 2.9 and 2.11, or powered auger for more difficult ground will reveal subsoil samples at various depths. Undisturbed samples should be removed for physical and chemical analysis. Level of ground water, if any, should be observed in the borehole and noted, as this could be a future excavation problem. A high water table during the summer months suggests surface flooding at less favourable times. Enquiries in the locality often prove more revealing than site tests and the local authority should have knowledge of problem areas. Geological maps will provide some insight into subsoil composition, but local knowledge is often the most reliable, revealing former rubbish tips, made-up ground and other unrecorded hazards. Physical analysis of the subsoil will reveal the type of strata to be found during excavation, its strength and bearing capacity. These are essential factors for foundation design, analysis of excavation costs and consideration of trench support systems. Chemical analysis is to reveal toxicity and contamination levels, with particular regard to sulphates which have an erosive effect on ordinary Portland cement mortars and concrete.

(*j*) *Connection of services to site*. Existing services through the site or over the site in the case of electricity or telecommunications will influence location of buildings. Consultation with the appropriate authorities may effect diversion of interfering services and extent of work and charges should be fully appreciated. Service pipes for foul and surface water drainage, drinking water, and gas must be located on or close to the site and expense and extent of work calculated for connections. Potential of electricity supply is to be determined, as use of three-phase high voltage for heavy plant items could be useful when scheduling plant.

(*k*) *Security*. Location of site with regard to potential trespass and materials theft is of paramount significance. Existing fencing should be reviewed and consultation with the local authority and police will determine the legal and advisable standards of site hoarding. Site security provision could extend to the use of compounds, barricades, night lighting and possibly security patrols.

4

(*l*) *Availability of labour*. Few builders employ all direct labour. The nature of the work lends itself to specialist sub-contracting and local firms engaged in the specialised aspects envisaged in the project should be listed for later reference. Details of local labour strength and availability should be determined in order to calculate demand on other labour resources and transport commitments. Amount of site labour will affect quantity of health and welfare facilities, including canteen, mess room, washroom and lavatories.

(*m*) *Local suppliers and builders' merchants*. Negotiation with materials suppliers and manufacturers will determine availability of materials, extent of credit and terms of settlement. Tool and plant hire firms should also be located and current charges reviewed.

Planning application

Application to the local authority for permission to develop land and construct buildings may follow satisfactory conclusion of the site feasibility study. However, to save time, where building permission has not already been granted, it should precede detailed site investigation. Permission is not always granted and a declined application will prevent further financial commitment to site analysis. Also, local authority planning meetings are held at intervals of several weeks and application following site investigation could waste useful time.

Applications are considered under several areas of legislation. The most significant Acts of Parliament are:

The Town and Country Planning Act 1971
The Public Health Act 1936
The Highways Act 1980
The Building Regulations 1976

The Town and Country Planning Act has effective control over the landscape, material appearance, layout and density of buildings. The Public Health Act prevents construction of buildings which could create a public nuisance from noise, refuse, air and water pollution. It also exercises control over sewage collection, disposal and treatment. The Highways Act controls location and construction of new footpaths and roads. Disruption and alteration to existing roads for formation of new accesses and connections to existing services must be effectively repaired and undertaken in a manner and standard acceptable to the local authority's highways department. The Building Regulations are a subsidiary instrument, not concerned with policy but with the material construction and design of building components. Their effect imposes strict control over location and suitability of materials and amenities.

Application procedure

Planning application is divided into two sections. The first is known as outline application and the second detailed application.

Outline application may contain only a plan of the site, a brief description of the proposed work and a certificate of ownership. One of four possible certificates apply, these are A, B, C or D. Certificate A states that the applicant is the estate owner in freehold or leasehold and that none of the land forms part of an agricultural holding. Certificate B states that the applicant has given notice to other persons having a freehold or leasehold interest in the land and certificates C and D are submitted when some or all persons entitled to notice cannot be located. This form of application provides the builder with a guide to site potential without incurring substantial expenditure on architect's plans and preparation of detailed information. If the site is particularly large or unusual, the local authority may request further information, including layout details and landscaping.

Full or detailed application is necessary when:

(a) outline permission has been granted;
(b) changing the use of an existing building; or
(c) extending an existing building or constructing an outbuilding.

In the latter case, this is only required if the proposed extension or outbuilding exceeds 70 cm^3 or 15 per cent of the original volume. For old buildings this is effective from 1948. Exceptions to the volume and percentage rule apply to proposals extending above the existing ridge or over the frontage line, (see Fig. 1.2). Front porches only require planning permission if they exceed 2 m^2 floor area, 3 m in height or are within 2 m of the boundary.

Detailed planning application contains:

(a) A plan of the site and adjacent area. This is the key plan and must be drawn to a minimum scale of 1 : 2500, with north direction indicated. See Fig. 1.1
(b) A site layout or block plan, drawn to a minimum scale of 1 : 1250, with north indicated.
(c) Detailed drawings, including elevations and plans to a minimum scale of 1 : 100. Sectional plans showing disposition of rooms should also be provided, preferably to a minimum scale of 1 : 50.
(d) Specification of materials. For small-scale house construction this is normally written on the drawing. When applied to large structures, it will be sufficiently detailed to form a separate document.
(e) Certificates of estate ownership – see outline application.

Planning applications are referred to the appropriate parish council and the plans are open to public view in the district council offices. A section of the local Press contains recent applications and comments or objections should be submitted within 21 days of publication.

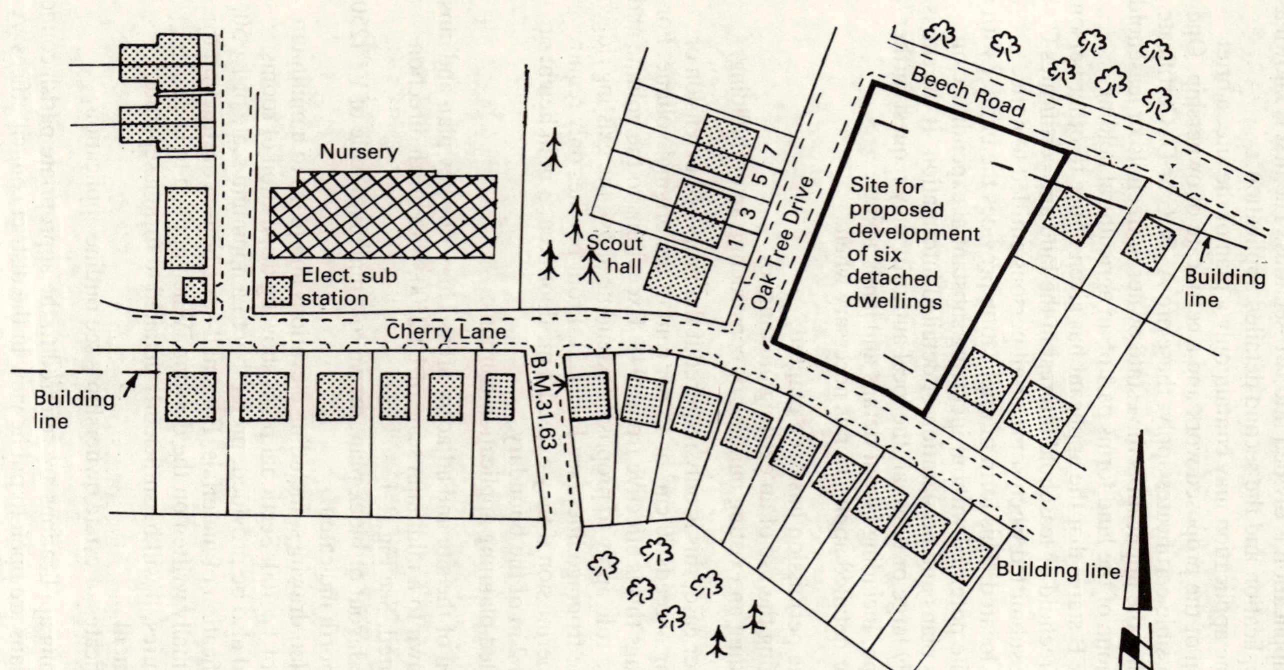

Fig. 1.1   Key plan 1 : 2500

Fees

Since 1 April 1981 the local authorities have administered a fee to partially cover the cost of processing planning applications. As an indication, a fee of £20 is required for enlargement, extension or alteration to existing buildings and for outline application a fee of £40 is required for each 1000 m^2 of site area up to a maximum of £1000. In other cases the fee is £40 for each proposed dwelling house up to a maximum of £2000.

Building Regulation application

Building Regulation approval is normally sought after planning permission has been granted. To save time it may be submitted before or at the same time, at the risk of resubmitting an amendment if planning permission is not straightforward. Building Regulation approval is necessary for:

(a) All new construction work
(b) Structural alterations to existing buildings
(c) Extensions to existing buildings
(d) Installation of fittings, e.g. sanitary ware
(e) Material change of use of a building, e.g. conversion of a house into flats.

Building Regulation applications are submitted to the building control section of the local authority and should contain the same information and plans required for detailed planning approval, with the exception of certificates. Additional requirements will be sectional details through significant parts of the structure, plus design calculations for any unusual elements of structure.

Building Regulation approval is confirmed with inspection notices provided for the builder to submit to the building control officer at appropriate stages of construction. The following require 24 hours' notice:

(a) Commencement of work
(b) Excavation of trenches for foundations
(c) Concrete in foundations
(d) Oversite hardcore laid
(e) Damp proof courses in position
(f)   Oversite concrete
(g) Soil and ventilating stack under test
(h) Foulwater drains laid in trench
(i) Surface water drains laid in trench
(j) Concrete over drains.

8

Seven days' notice is required for:

(k) Foulwater drain trenches backfilled and under test
(l) Completion or occupation of the structure.

This list is no limit to the influence of the building control officer.
Additional random visits can be expected to ensure compliance with
the approved plans.

Fees

A fee is also required for Building Regulation approval and building
control services. The amount is approximately 1 per cent of the value
of work subject to approval, and one-third of the fee must be submitted
with the application. The remainder is paid before final inspection
or completion.

Setting out

Setting out follows site levelling and stripping of the topsoil. It is the
process of accurately locating and marking out the position of

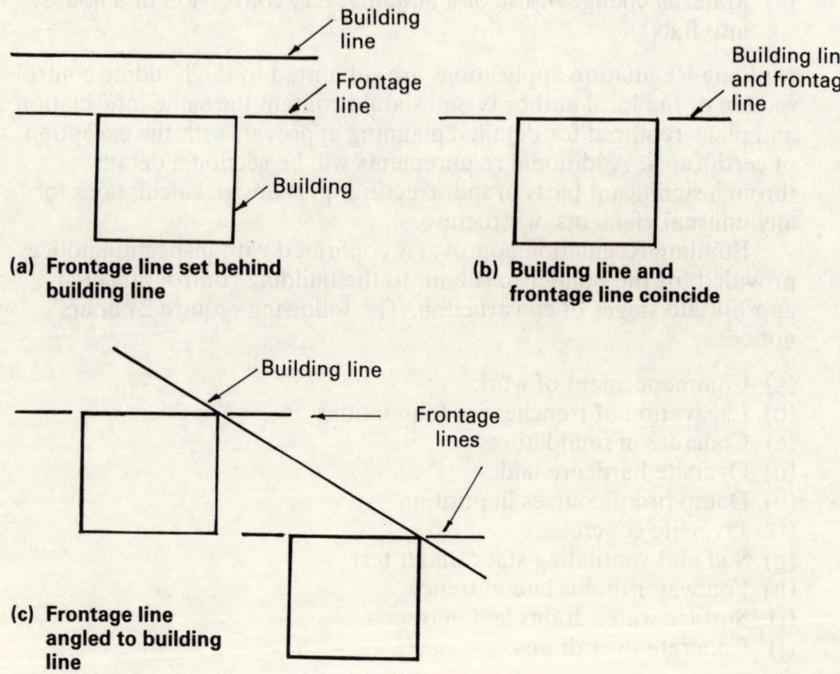

Fig. 1.2   Location of frontage lines

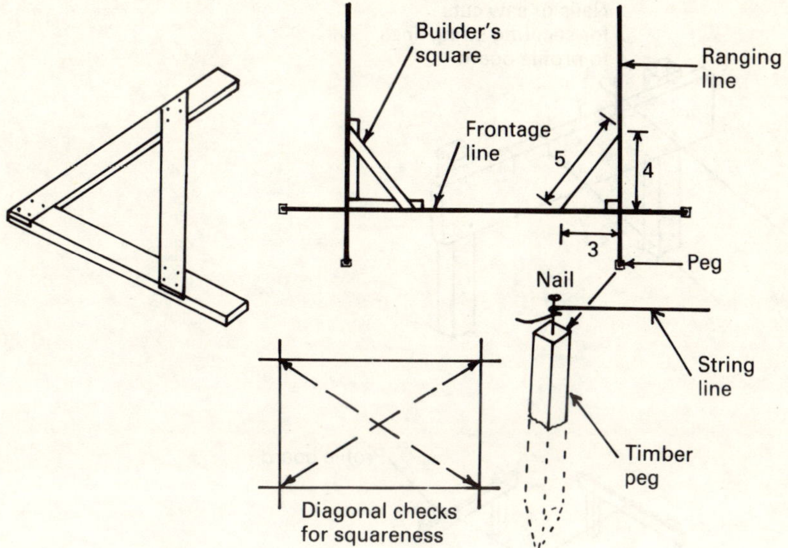

Fig. 1.3 Setting out corners

foundation trenches and walls. A frontage line is obtained from block and key plans to determine the position of the building relative to the footpath or highway. This may coincide with the building line and the relationship is shown in Fig. 1.2(a–c). The building line represents a line defined by the local authority as the closest position which a building is permitted to the road centre line or kerb line.

The frontage line is established by extending a string line between well-driven stout timber pegs. Accuracy is ensured by using a nail to position the string line as shown in Fig. 1.3. The front corners or quoins are positioned with perpendicular ranging lines to represent the outer line of the side walls. This is illustrated in Fig. 1.3, which also shows two methods of ensuring a right angle between lines. The simplest method uses a large timber set square or builder's square and the other is an application of Pythagoras' theorem. This provides a right-angle in a triangle having sides in the ratio of 3 : 4 : 5. These methods are of limited accuracy and not recommended for large buildings where the use of instruments are more suitable. Chapter 7 of *Site Surveying and Levelling*, Level 2, by H. Rawlinson (Longman Technician Series), provides details of the use of both site square and theodolite in setting out large areas.

Offsets and additions are positioned similarly and when the outline of the building is finally checked and found accurate, profile boards are erected clear of the string lines. These are shown in Fig. 1.4 as horizontal boards nailed to stout pegs firmly driven into the ground. The original string lines are transferred to the board and secured by

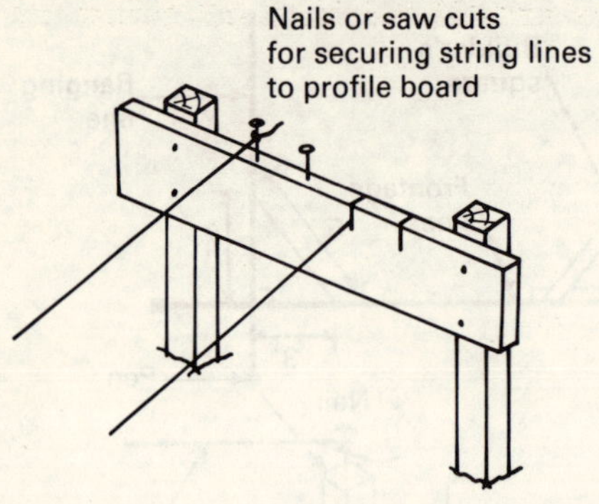

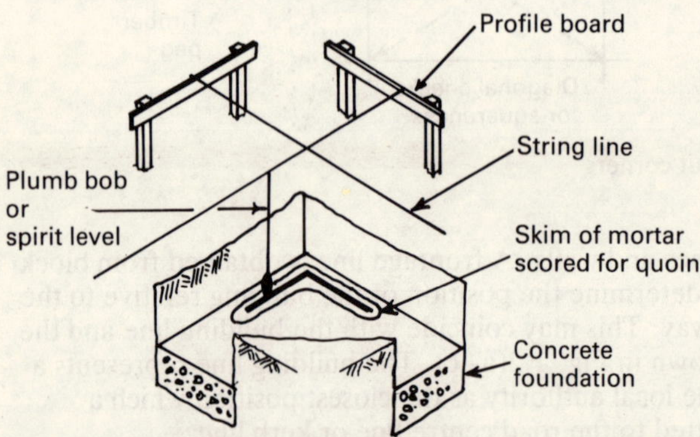

Fig. 1.4 Use of profile boards

nail or saw cut. The inner face of the wall and trench excavation lines are measured from setting out lines and also located with saw cuts or nails. String lines are suspended between the profiles to represent the trench width and a line of sand or cement is run along the ground beneath the string to provide guidance for the excavating bucket when the string is removed.

Following excavation and concreting, string lines are re-established on the profile boards for locating the wall. These positions are transferred to the concrete by plumb bob or level as shown in Fig. 1.4, and a thin skim of mortar is scored to represent the corner. After sub-structural brickwork has commenced, the profile boards are removed as bricklayers will suspend string lines from the quoins to ensure alignment of the wall.

Chapter 2

Excavation work on construction sites

Substructure

The substructure is the part of any building which is below the natural or artificial level of surrounding ground. It has the function of transferring superstructural dead and imposed loads to the natural foundation or subsoil receiving the building. (See Fig. 2.1 for the relationship between substructure and ground conditions.)

Removal of vegetable soil

This is a primary function, undertaken before construction commences. Clearance of vegetable soil or topsoil from the construction area is mandatory and specified in Building Regulation C2 par. (I). 'The site . . . shall be cleared of turf and other vegetable matter.'

Vegetable soil is soft, easily compressed and contains plant life, properties excluding its use as a basis for construction. It is found in variable depths, but not usually in excess of 300 mm. In some parts of the country excellent growing matter extends well beyond this depth; in these situations conventional foundations and excavation techniques are unsuitable. It is stripped and retained to one side of the site for re-use when landscaping the finished product. If not required, it is of value as a supplement to existing gardens containing poorer quality topsoil.

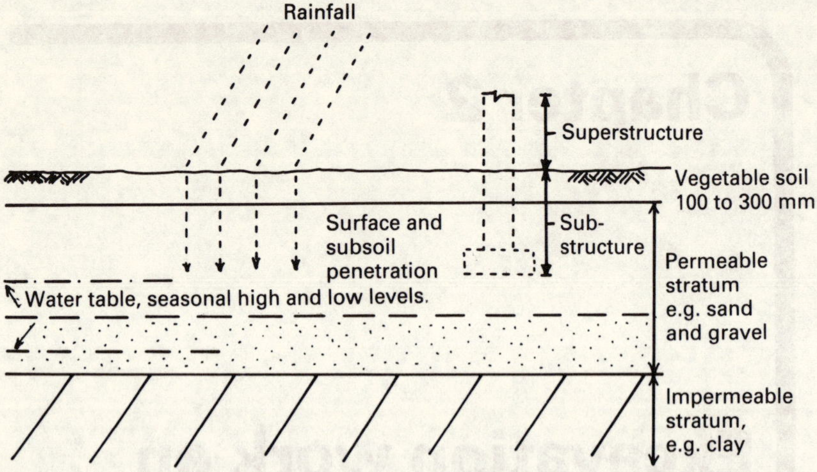

Fig. 2.1   Subsoil and water table

Types of excavation

Excavation is that part of the construction process where greatest
variables occur, as subsoil content and condition varies even within a
relatively close area. Before selection of excavation technique, site
investigation to obtain knowledge of the subsoil is essential, in addition
to location of underground services, mine shafts, wells and the natural
level of water. These factors are frequently undetected on maps and
can generate considerable difficulties and unforeseen costs.

Types of excavation can be classified as:

1. Hand tools
2. Mechanical plant, used for:
    (a) Stripping vegetable soil
    (b) Reducing levels. Cut and fill
    (c) Trench excavation
    (d) Pit excavation

1. Hand tools

Excavation by hand tools is now virtually obsolete, except for trimming
to mechanical excavations. The other exceptions are:

1. Where the volume of excavation is minimal, e.g. for a small
   extension
2. In situations where mechanical plant accessibility is restricted
3. Where, on very rare occasions, construction is on the site of ancient
   remains and careful handling is specified.

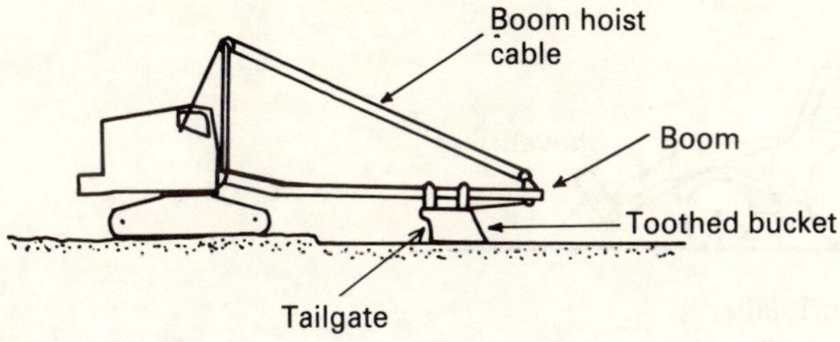

Fig. 2.2 Skimmer

2. Mechanical plant

(a) *Stripping vegetable soil*

A variety of machinery may be employed for this purpose, selection
depending on the nature and volume of the topsoil and site conditions.
Graders and skimmers shown in Figs 2.2 and 2.3 have been popular for
many years, but for most housing sites the face shovel attachment to
an agricultural type tractor of the type shown in Fig. 2.4 is quite
adequate. Graders are very versatile, having a forward facing blade
which can be raised, lowered and angled to cut embankments and
ditches. Skimmers have a hinged door, forward-facing bucket
travelling along a boom, and are most suitable for highway and other
large-scale surface excavation. The bucket is emptied by raising the
boom and opening the door.

Bulldozers are ideal for land clearance and demolition. With
bucket lowered and angled they function well, stripping off topsoil,
levelling the site and spreading surplus material. This is illustrated in
Fig. 2.5 on sloping sites where the excavation involves cutting into the
slope and filling below.

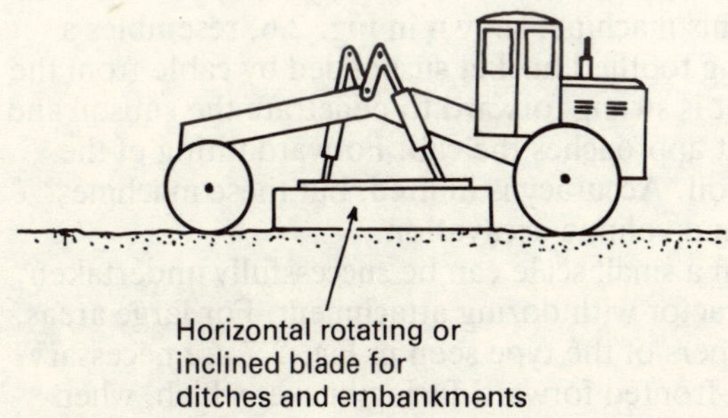

Fig. 2.3 Grader

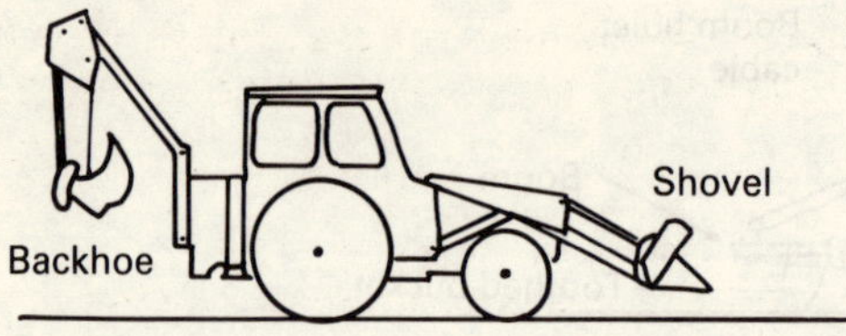

Fig. 2.4 Excavator/Loader

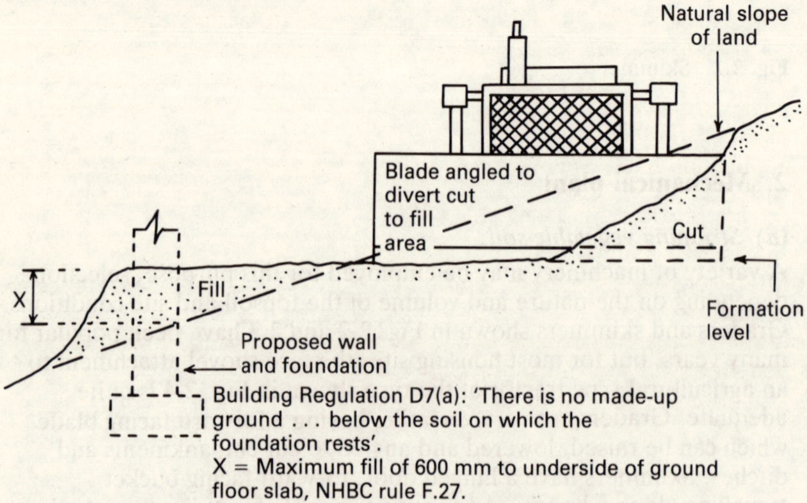

Fig. 2.5 'Cut and Fill' using a bulldozer to provide a level site

(b) *Reducing levels*

Reducing levels is a method of lowering the ground level to sufficient depth for semi-basement construction or by application of 'cut and fill' techniques to provide a level base for construction on sloping housing sites.

For larger basements a dragline may be employed, operating from above the excavation. This machine, shown in Fig. 2.6, resembles a crane but has a cab-facing toothed bucket suspended by cable from the lattice boom. The bucket is swung forward to penetrate the subsoil and dragged back, filling as it approaches the cab. Forward tilting of the bucket discharges the spoil. Accuracy is limited, but these machines are ideal for large area and volume excavation.

Reduced levelling on a small scale can be successfully undertaken by wheeled or crawler tractor with dozing attachment. For large areas, towed or motorised scrapers of the type seen in Fig. 2.7 are necessary. They have a large, open-fronted forward-facing bucket which, when lowered, excavates in shallow cuts. When filled, the bucket is raised, taken to a tipping area where a hydraulic ram displaces the contents as the machine is driven forwards. These machines are particularly useful for levelling to highways and can produce large-scale excavation with precise formation levels.

(c) *Trench excavation*

Trenchers or ditchers as shown in Fig. 2.8 are an alternative to the popular backhoe, but are now a rare sight in construction work. They consist of an endless chain of excavating buckets which rotate to remove the subsoil and deposit it to the sides of excavation.

The backacter has a toothed bucket with jib and boom reaching out and excavating towards the cab. Spoil is deposited by lifting the bucket up and away from the excavation. Most machines are capable of excavating trenches well in excess of 5 m, and also have an application to basement and ditch excavation. In trenching it functions with crawler tracks both sides of the trench and driver's cab spanning the line of excavation.

The wheeled agricultural tractor has become a very popular base for excavating equipment. With backacter and face shovel attachments at opposing ends, it is ideal for light excavation to foundation and drain trenches. These are most suited to housing sites and have the benefit of self transportation between sites.

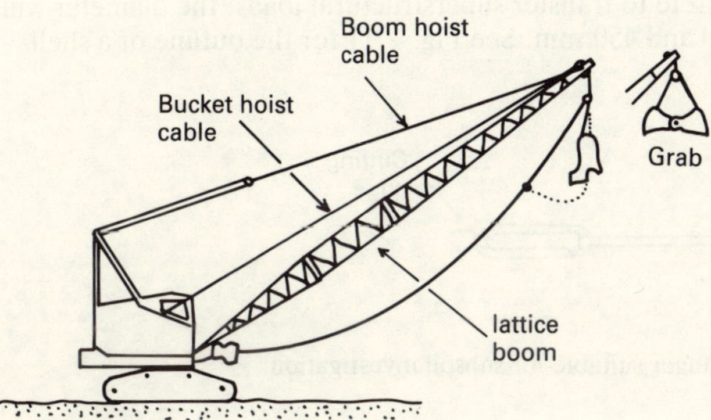

Fig. 2.6 Dragline/Grabcrane

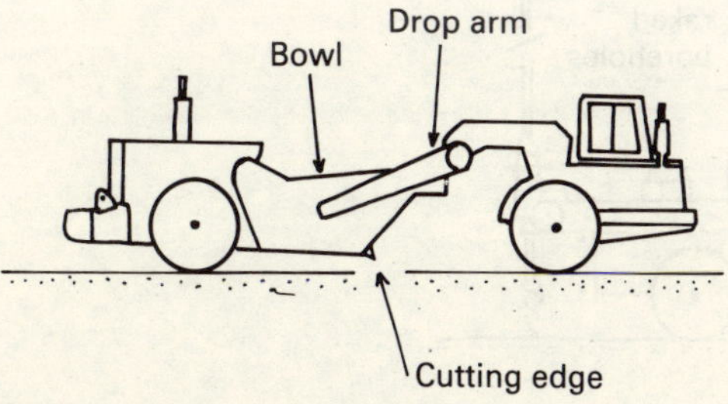

Fig. 2.7 Double-engined scraper

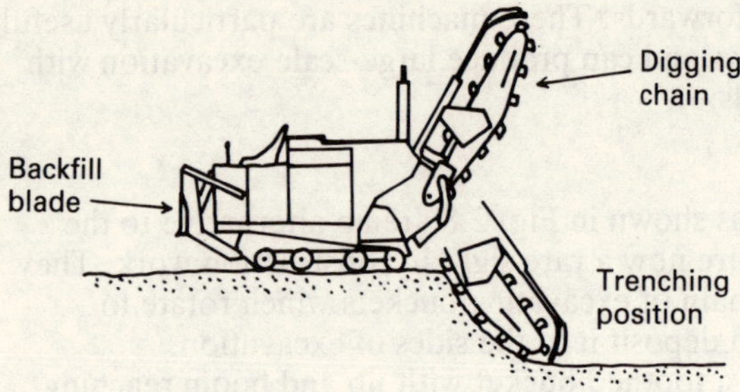

Fig. 2.8 Tracked trencher

(d) Pit and augered excavation

Augered or bored excavations are used initially as a means of
ascertaining the composition and nature of the subsoil. Shallow
boreholes can frequently be undertaken manually with the hand auger
shown in Fig. 2.9. For granular subsoils and deep borings, the
lorry-mounted auger illustrated in Fig. 2.10 is ideal but where soft
subsoils exist tripod suspended shell augers are preferred. For subsoil
analysis small diameter boreholes are sufficient, but as a foundation,
filled with concrete to transfer superstructural loads, the diameter will
be between 250 and 450 mm. See Fig. 2.11 for the outline of a shell
auger rig.

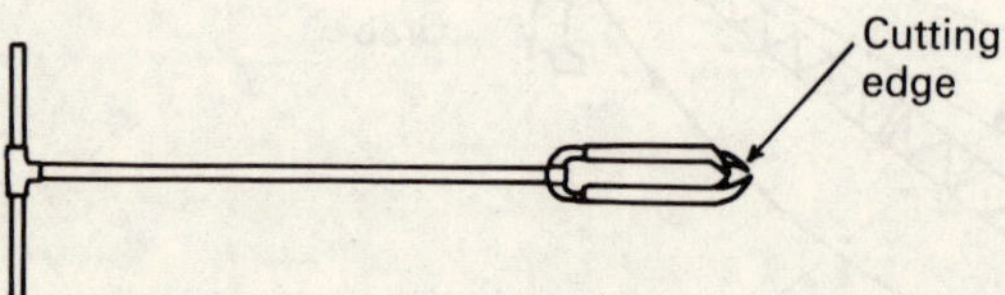

Fig. 2.9   Hand auger suitable for subsoil investigation

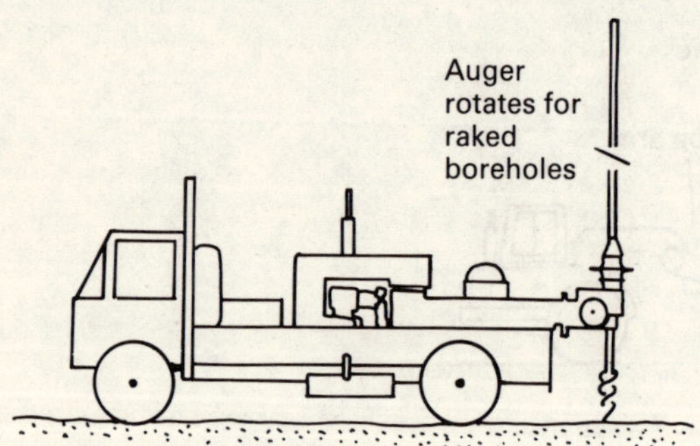

Fig. 2.10   Lorry-mounter auger

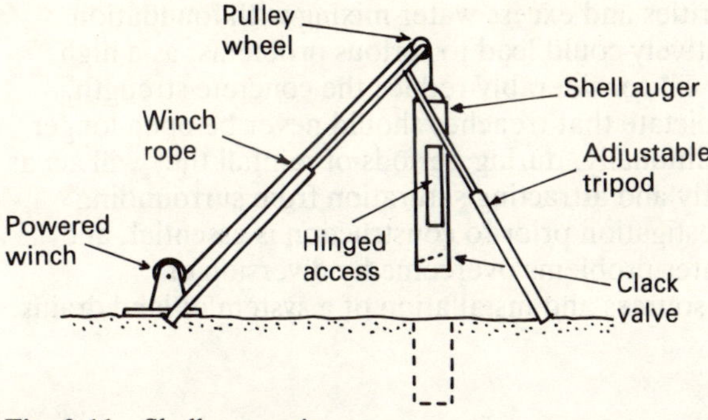

Fig. 2.11 Shell auger rig

Another application of borehole excavation is for land drainage of large flat areas. Here, borings are filled with a drainage medium such as sand which will effectively reduce surface flooding by drawing the water down to naturally draining stratum.

Large voids in granular subsoils may be created by the grab or clamshell indicated in Fig. 2.6. These have an open-top bucket, divided into two vertical sections and suspended from a crane jib. To penetrate, the bucket is dropped with jaws open. Spoil is removed and deposited by closing the jaws, raising the bucket and reopening it at a convenient point of discharge.

The need for removal of water from excavation

The subsoil will always contain a quantity of water, the amount and level varying depending on the nature and composition of underlying stratum. Surface saturation, after periods of excessive rainfall, can create problems during and after building, therefore means for its rapid removal must be incorporated into the construction programme.

The natural level of water saturation of the subsoil is referred to as the water table, and its height varies according to seasonal extremes. Following a reasonably dry summer it will be low when compared with the level after the winter season. Before installation of piped drinking water to dwellings, the water table would have been the source of well supplies.

Excavation into the water table is undesirable as this hinders machinery and provides a very unpleasant work situation. If possible the advantages of summer excavation should be exploited. The presence of water will undermine the sides of excavation and cause subsoil spillage, producing an irregular and uneven trench. The need for trench support will be incurred in addition to excavation for sumps to attract a large volume of water for pumped extraction. Dry excavation is essential for safety and ease of working in addition to the need to avoid impurities and excess water mixing with foundation concrete. This effectively could lead to serious problems, as a high water/cement ratio will considerably reduce the concrete strength.

Safety reasons dictate that trenches should never be open longer than necessary. Additionally, during periods of rainfall they will act as a ditch, filling directly and attracting saturation from surrounding subsoil. Subsoil investigation prior to construction is essential, and potential ground water problems overcome by diversion of underground water sources and installation of a system of land drains.

Chapter 3

Work below ground

Safety in open excavations

Excavation to form foundation and drain trenches is one of the first operations to follow site preparation. Few excavations are self-supporting, and the duration and depth of excavation will significantly affect the stability of exposed subsoil.

The 1961 Construction (General provisions) Regulations made under the 1961 Factories Act, but now under the direction of the Health and Safety at Work, etc. Act 1974, ensure safe working conditions in excavation and earthworks. PART IV, section 8(1) specifies 'An adequate supply of timber . . . used to prevent . . . danger to any person employed . . . adjacent to any excavation'. This requirement is excepted if the depth of excavation is within 1.2 m, or the nature and slope of the subsoil are sufficient to render the excavation stable without secondary support. A further exception applies to excavations where operatives are engaged in the installation of timbering. In these circumstances, appropriate measures must be taken to suit the situation.

During every working day a person of sufficient responsibility and competence must inspect all parts of an excavation to ensure safe working conditions. All supporting materials must be of sufficient strength for their function, and maintained free from defect. Sides of excavations must be sufficiently strutted, braced and secured to prevent accidental movement.

Excavations in excess of 2 m shall be provided with a suitable

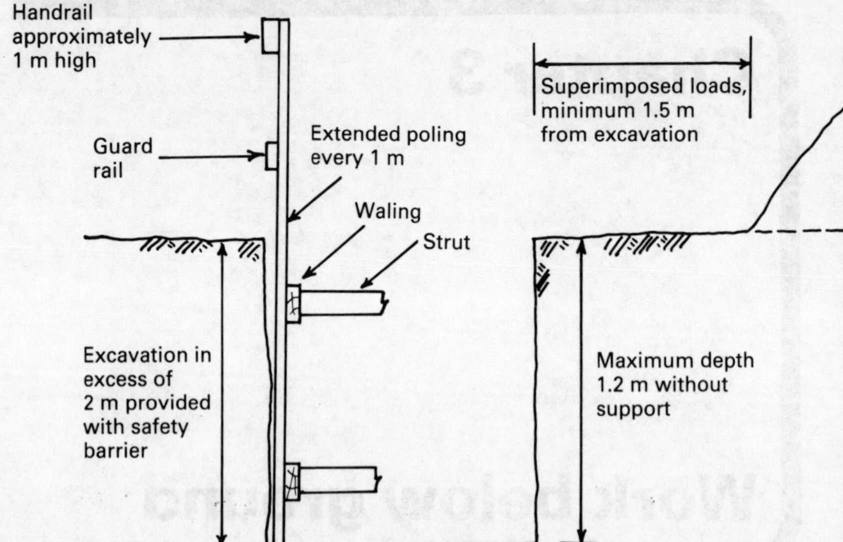

Fig. 3.1 Safety factors in trench excavation

fence or barrier, possibly incorporated within the trench support system or provided as close as possible to the edge. This does not apply during placement and removal of persons, plant, materials and equipment. Edges of all excavations must be kept clear of superimposed loads likely to cause collapse of the trench sides. This includes spoil from the excavations, plant and materials which could endanger persons below the surface. See Fig. 3.1 for the application of these requirements.

Support to open excavation

British Standard 6031 : 1981 provides a guide to basic methods of trench excavation and construction of temporary support to excavation sides. A trench is defined as 'An excavation whose length greatly exceeds its width and which has either vertical sides capable of being supported by strutting from side to side or battered sides requiring no support.' (See Fig. 3.2.)

Types of trench are defined by their depth and use:
Shallow trench – up to 1.5 m, used for services and strip foundations.
Medium trench – between 1.5 and 6.0 m, used for pipelines and sewers.
Deep trench – over 6.0 m, for various uses.

Support to excavations is known in general terms as timbering or planking and strutting, but in practice components are not restricted to

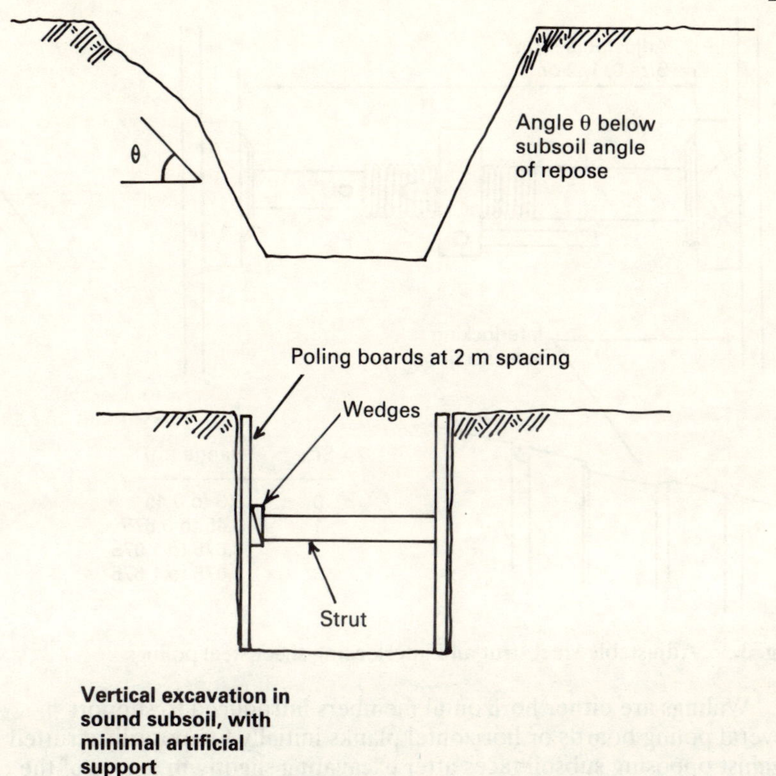

Fig. 3.2 Self-supporting and artificially supported trenches

the use of timber. Inter-locking sheet steel sections and adjustable steel struts of the types shown in Fig. 3.3 are frequently used, although the former are normally applied to deep, large or water-saturated excavations. The components of timbering are basically poling boards, waling and struts.

Poling boards are used facing the exposed subsoil in shallow trenches where the excavation will be self-supporting for sufficient time to enable placing. Dimensions vary, but normally range from 1 to 1.5 m in length, possibly projecting above ground level as support to a guard rail. Widths are between 175 and 225 mm, and thickness from about 35 to 50 mm.

Note: Accurate determination of timbering component size is possible using well-established formulae based on computation of pressures and subsoil analysis. This is beyond the scope of this level and in practice rarely applied, as the temporary nature of this work does not justify the time.

Selection of suitable timber is largely left to a person sufficiently knowledgeable and experienced in this type of work.

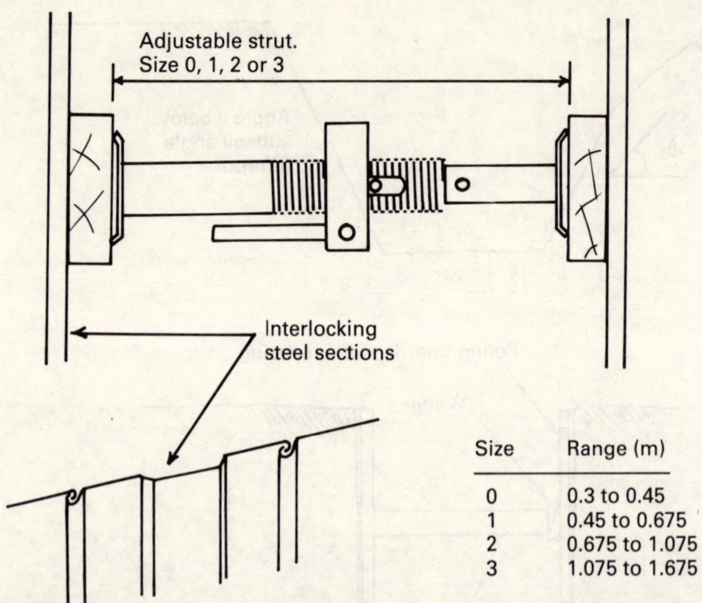

Size	Range (m)
0	0.3 to 0.45
1	0.45 to 0.675
2	0.675 to 1.075
3	1.075 to 1.675

Fig. 3.3   Adjustable steel strut and interlocking sheet steel polings

Walings are either horizontal members introduced to support several poling boards or horizontal planks initially temporarily strutted against opposing subsoil faces after excavating slightly in excess of the board width. When reaching about 1 m in depth, soldiers are positioned against the planking, and struts permanently located. This technique is favourable in saturated or loose subsoils where timbering is essential with the advancing excavation. Trenches of considerable depth and uniform width requiring constant lining will also benefit from the use of waling boards. Used as support to polings, walings range from 100 mm × 75 mm to 100 mm × 175 mm; as sheeting or planking from 225 mm × 38 mm to 225 mm × 75 mm, and in both situations between about 2.5 to 5.4 m long.

Struts are horizontal members designed to resist subsoil thrust and earth pressure from opposing sides of the excavation. As timber sections they are cut to the required length and driven down between polings or walings. Alternatively, and where excavated trench sides are parallel, pairs of overlapping wedges (folding wedges) provide the strut with compressive force to uphold the earth-facing timbers. Square section timber is suitable of 75 mm or 100 mm for short spans, but for large openings with no intermediate support considerably larger sections must be employed. Where large spans exist, it often proves more economical to use steel screw struts, and because of their flexibility and re-useable value these are now frequently employed for narrow excavations. Examples indicating the use of trench timbering are shown in Fig. 3.4.

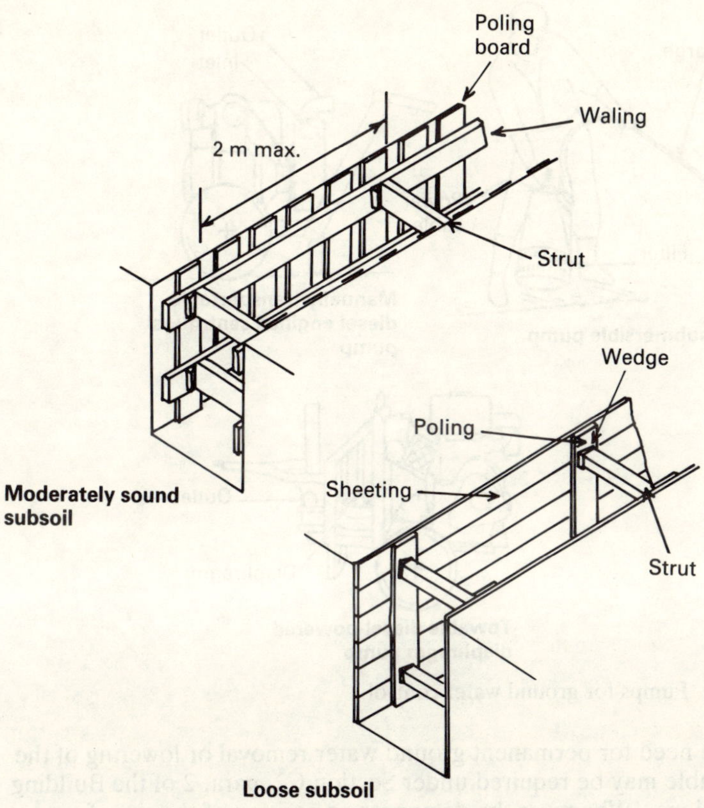

Fig. 3.4 Simple trench timbering systems

Provision of workable conditions by extraction of ground water

Methods for removing ground water in ground excavations are
determined by the desire for a temporary or permanent effect.
Temporary methods are simpler and involve excavating one or
preferably two small pits or sumps at positions within or close to the
excavation. Pumping from the sumps must be continuous if the water
level is to remain below the level of excavation, and type of pump used
will depend on the quantity of water and mud content. If the water is
muddy or gritty a fuel-oil-powered diaphragm pump of limited capacity
is most suited. For clear water and greater capacity, the centrifugal
pump is preferred, but clean conditions are unlikely in excavation
work. Electric centrifugal pumps for ground-level use or in submersible
form are gaining popularity, but are not always convenient during the
early stages of construction – when electricity may not be available.
The impellor blades which pressurise the water are made open with
wide passages for handling muddy mater containing solids or enclosed
with narrower passages for clear water.

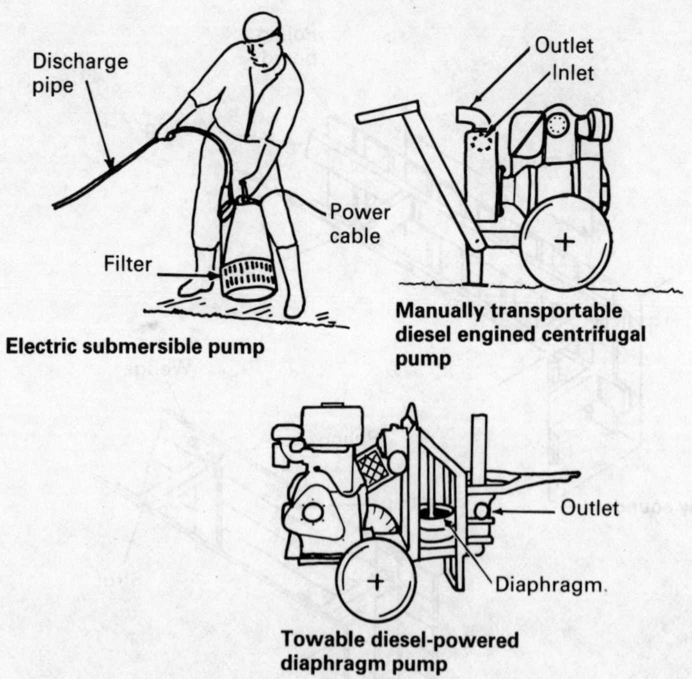

Fig. 3.5 Pumps for ground water control

The need for permanent ground water removal or lowering of the water table may be required under Section C2, para. 2 of the Building Regulations, 'Whenever the dampness or position of the site of a building renders it necessary, the subsoil of the site shall be effectively drained . . .' This is essential to eliminate ground water pressures and to reduce the effect of frost heave on the sub-structure. It also removes the possibility of water movement segregating the components of the subsoil, thereby reducing its bearing capacity. In addition to the Building Regulations, the National House Builder's Council require water-logged areas within 3 metres of a proposed dwelling house to receive suitable precautions to prevent the saturation affecting the structure – rule S18(a).

The simplest method for satisfying these requirements is a French or rubble drain of the type shown in Fig. 3.6. These are trenches excavated to sufficient depth (rarely over 1.2 m) and to a gradient, filled with graded granular material. They function as both carrier and filter, and if filled to ground level also function as a surface water cut off. As a carrier and filter, coarse granular fill or rubble occupies most of the trench, the remainder having a brushwood or straw filter superimposed with normal backfill and turf. The life of these drains is limited to the filter effectiveness, and eventually they become clogged with fine subsoil particles.

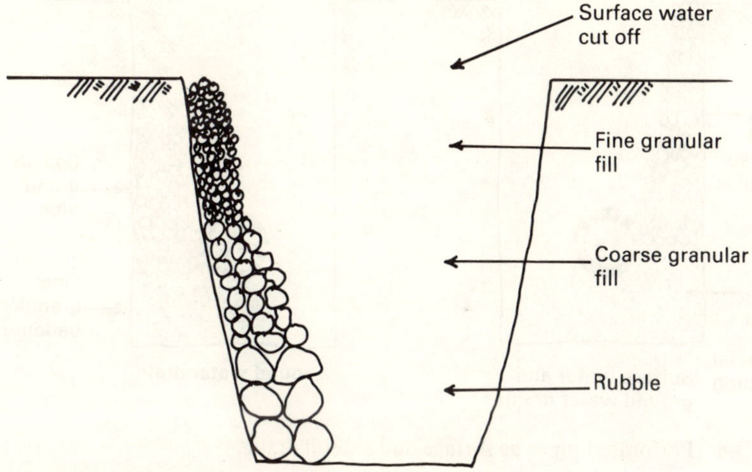

Fig. 3.6 French or rubble drain

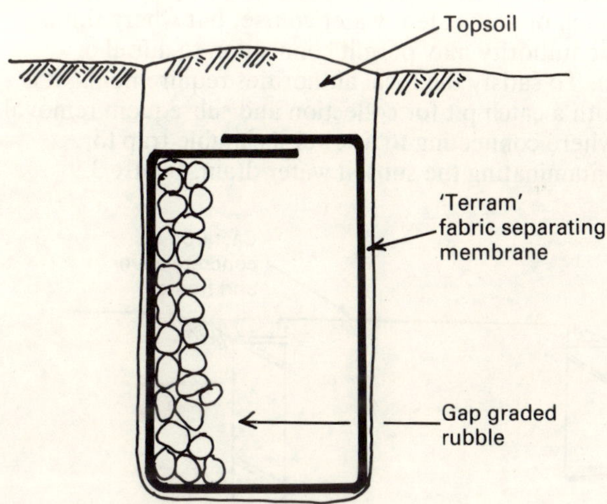

Fig. 3.7 Improved rubble drain

Figure 3.7 shows a recent improvement in the use of rubble trenches, and includes the use of a fabric membrane surrounding the fill material, functioning as a filter and water retainer. The advantages include the use of a lower quality gap graded rubble as drainage media, thus improving flow conditions and economising in trench size and quantities of filling.

A less recent innovation shown in Fig. 3.8 utilises porous or perforated pipes laid close to the bottom of a granular filled trench. These collector pipes are more reliable than rubble drains as they offer

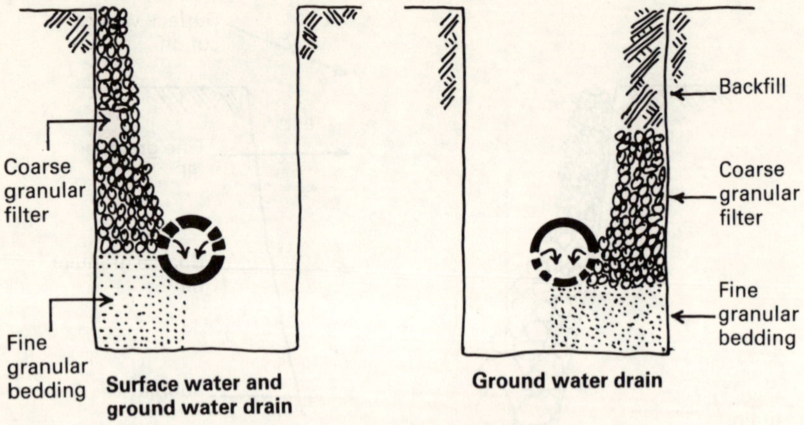

Fig. 3.8 Perforated pipes as surface and subsoil drains

a permanent drainage channel, but they too eventually suffer from clogging and obliteration of their external surface. The pipe run is best discharged into a ditch or convenient water course, but where this is impossible the local authority may permit connection to a foul or surface water drain. To satisfy the local authorities requirements, it is usual to provide both a catch pit for collection and subsequent removal of slit or grit and where connecting to a sewer a suitable trap to prevent foul air contaminating the subsoil water drain. Figure 3.9

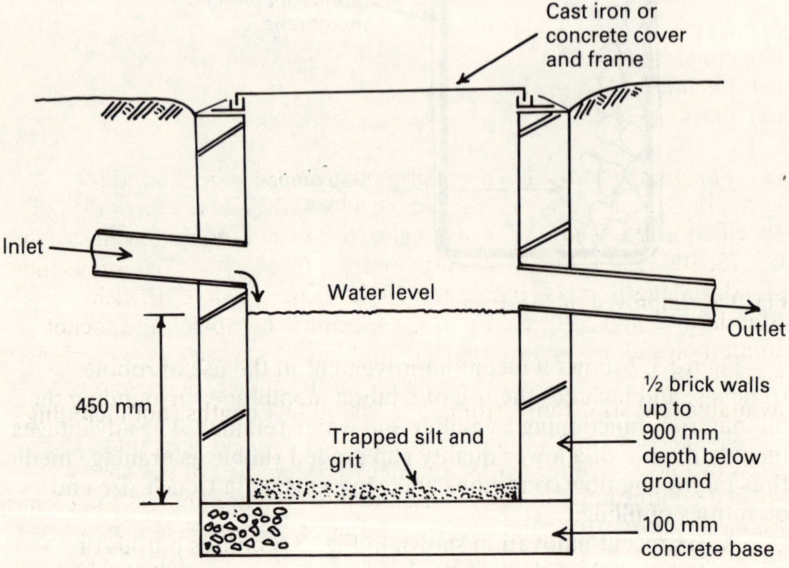

Fig. 3.9 Silt catch pit

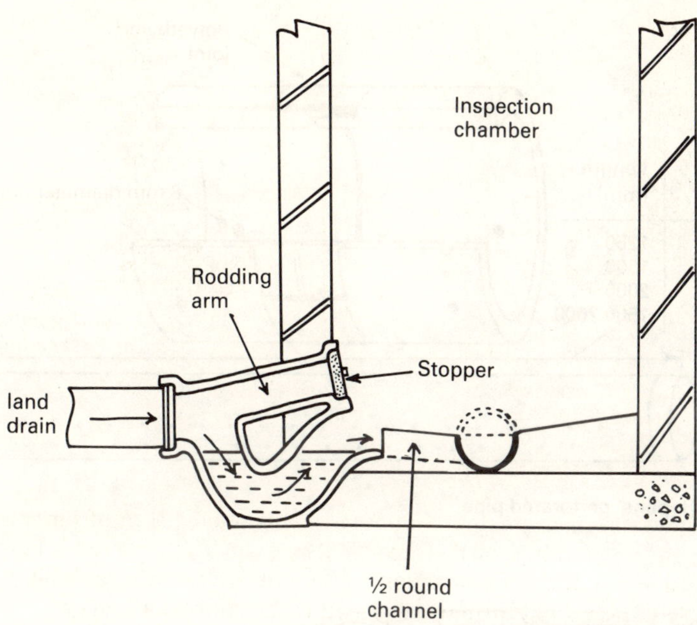

Fig. 3.10   Use of a reverse acting interceptor trap

shows construction of a suitable catch pit and Fig. 3.10 a reverse acting
interceptor trap for use where a land drain connects to a sewer.
   Materials for subsoil water drains vary, the most popular include:

(a) clayware (Fig. 3.11)
(b) concrete (Fig. 3.12)
(c) pitch fibre (Fig. 3.13)
(d) plastic (Fig. 3.14)

(a) Porous clayware field drain pipes. These have been used for
agricultural purposes for hundreds of years. In modern form they are
specified in BS 1196 : 1971, as unglazed butt jointed and available in
65, 75, 100, 150, 225 and 300 mm internal diameters – 300 mm is the
standard length, but 450 mm and 600 mm are possible. British
Standard 65 and 540 : Part I : 1971 specifies clay spigot and socket
jointed pipes for general drainage, and includes a section on perforated
subsoil drain pipes. Pipes of many different internal diameters are
available and sizes range from 75 to 900 mm. Lengths (not including
the socket) are variable and range between 300 and 1500 mm. The
perforations are contained within one 180° segment of the barrel,
spaced equidistantly in rows along the pipe and are not less than 3 mm
diameter, nor greater than 15 mm diameter. To ensure effectiveness
the total area of holes must not be less than 1000 mm^2 per 1 m length
of pipe.

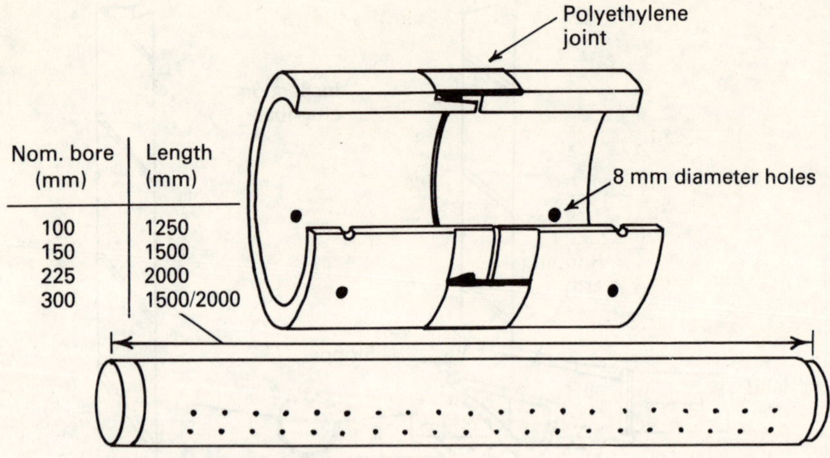

Nom. bore (mm)	Length (mm)
100	1250
150	1500
225	2000
300	1500/2000

'Hepline' perforated pipe

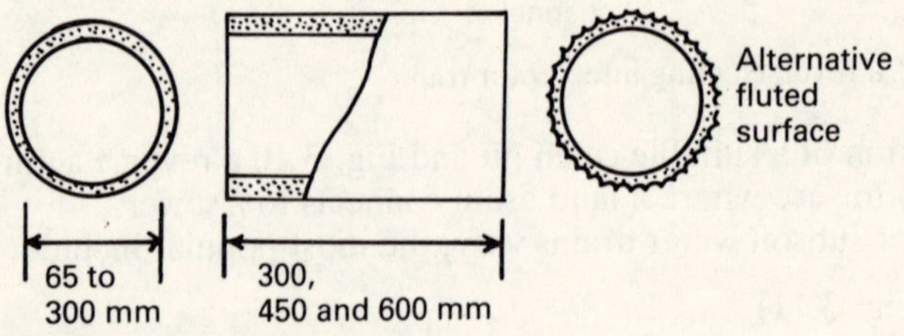

Porous clay pipes to BS 1196

Fig. 3.11 Clayware subsoil drain pipes

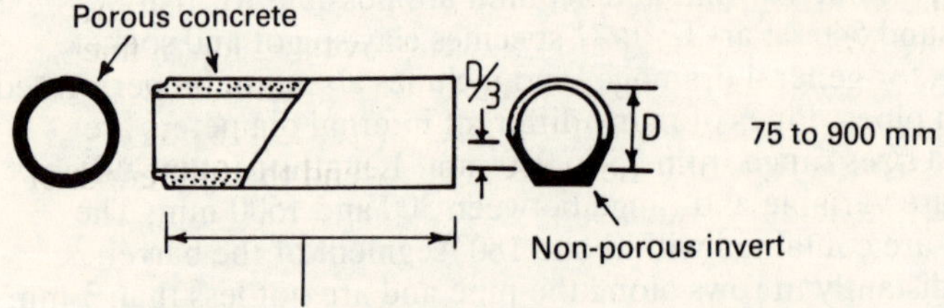

75 mm diameter, 0.3 m min–0.6 m max.
Over 75 mm, 0.45 m min–1.0 m max.

Fig. 3.12 Porous concrete subsoil drain pipe

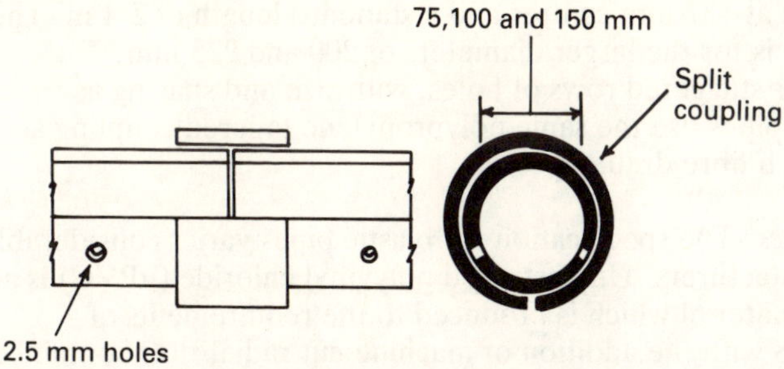

Fig. 3.13 Butt-jointed pitch fibre perforated pipe

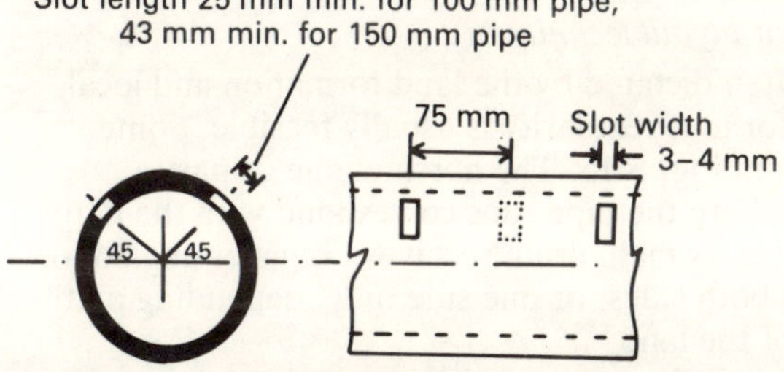

Fig. 3.14 uPVC slotted pipe

A variation in vitrified clayware, which still satisfies BS 65 and 540, is the 'Hepline' land drainage system using push-fit flexible polyethylene sleeve joints. These pipes are available in 100, 150, 225 and 300 mm internal diameters, with lengths of 1250, 1500 and 1600 mm respectively. The pipe length is perforated with 8 mm diameter holes provided in rows contained within an arc of 160°.

(*b*) *Concrete porous pipes.* Used in subsoil drainage, these are specified in BS 1194 : 1969. Several sizes are available and internal diameters range from 75 to 900 mm. Lengths for 75 mm pipes are between 300 and 600 mm, and for pipes over 100 mm, between 450 mm and 1 m. Connection is by ogee or rebated joint contained within the pipe wall thickness. Porosity occurs by omission of fine aggregates or by using an aggregate which encourages the passage of water. Pipes can be made with a non-porous invert to provide a channel to assist the flow of water; this must extend the full length of the pipe and to a height of approximately one third the internal diameter.

(*c*) *Pitch fibre subsoil drain pipes.* These are divided into two categories. Firstly, with plain ends and split 'C' type couplings and internal diameters of 75, 100 and 150 mm. These have two rows of

12.5 mm holes at 127 mm centres and a standard length of 2.4 m. The other category is for the larger diameters of 200 and 225 mm, containing four staggered rows of holes, with size and spacing as before. These pipes use the same polypropylene tapered coupling as for normal pitch fibre drain pipes.

(*d*) *Plastic pipes*. The specification for plastic pipes varies considerably between manufacturers. Unplasticised polyvinyl chloride (uPVC) is a popular pipe material which is produced to the requirements of BS 3505 : 1968 with the addition of machine-cut radial slots contained within 45° of the pipe crown. Internal diameters include 100, 150, 200, 225, 250 and 300 mm and lengths are 3, 6 and 9 m, joined by a push-fit spigot and socket.

Subsoil drain systems or layout techniques

Layout patterns are often dictated by the land formation and local conditions, but some form of regularity is usually feasible. Some possibilities are shown in Fig. 3.15. The herringbone or part herringbone is useful where the pipe runs correspond with the natural slopes of an inclined site. A main drain receives branches angled to encourage water from both sides, or one side only, depending on the size of site and slope of the land.

The gridiron pattern is the most regular layout, having a main drain to one side of the site, clear of construction work with tributory drains connected at right angles. Basement construction or renovation work to existing basements often justifies inclusion of a subsoil drain around the perimeter of the building to reduce or eliminate the effects of water pressure. This is known as the moat system and has the effect

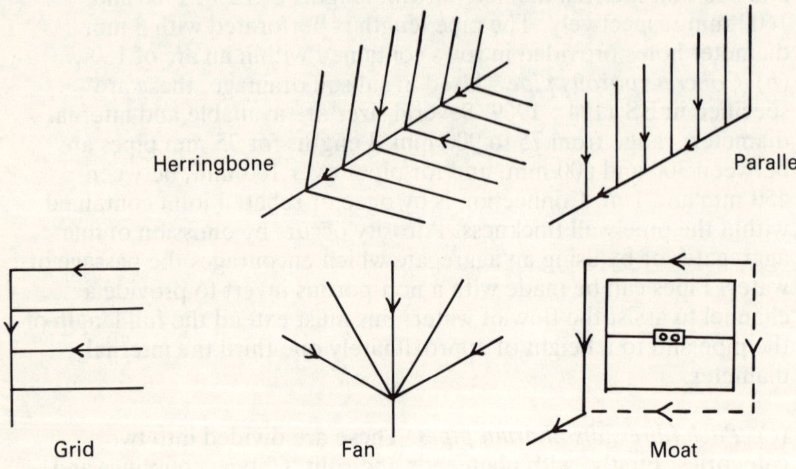

Fig. 3.15   Systems of land drainage

of intersecting ground water flow to a structure as well as reducing the water table.

Subsoil drains are often used in a single line to cut off or intersect the flow of ground water to a site or as a modified ditch, beside highways where gullies and surface water drains are not provided. Here they prevent rising ground water causing surface flooding and provide a rapid run off for rainwater diverted over the sides of the road.

Chapter 4

Drainage

Rainwater conveyance

Section 37 of the 1936 Public Health Act enables local authorities to insist on satisfactory provision of rainwater conveyance from the roof areas of all new and existing buildings.

The 1976 Building Regulations, parts N8 and N9(1), are more precise in accounting for the correct design and installation of suitable rainwater goods, and N2(1) defines a rainwater pipe as 'a pipe (not being a drain) which conveys only rainwater'. Where the rainwater pipe enters a foul water drain, regulation N13 requires the rainwater to be 'intercepted by a suitable trap, having a seal not less than 50 mm in depth'. Ideally the rainwater pipe should connect below the level of the grating of a gulley. See Fig. 4.1 for an example of a back inlet gulley which is ideal for this situation. If the drain is solely for surface water, the rainwater pipe may connect direct without a gulley, the rainwater shoe shown in Fig. 4.2 being most suitable as it provides access in the event of blockage. This does not preclude the use of a gulley; in fact, they are often preferred because the trap will stop leaves and other debris caught by the gutter creating a blockage in a less accessible position.

Building Regulations N8 and N9(1) are concerned with guttering and external downpipes respectively. Each regulation has four identical sections, (a), (b), (c) and (d). N8 continues with two further sections, (e) and (f).

Section (a) requires the system to be 'of adequate size for its

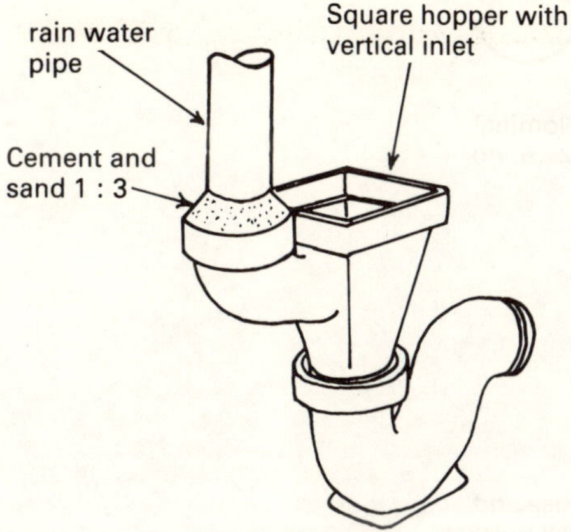

Fig. 4.1 Back inlet gulley in clayware

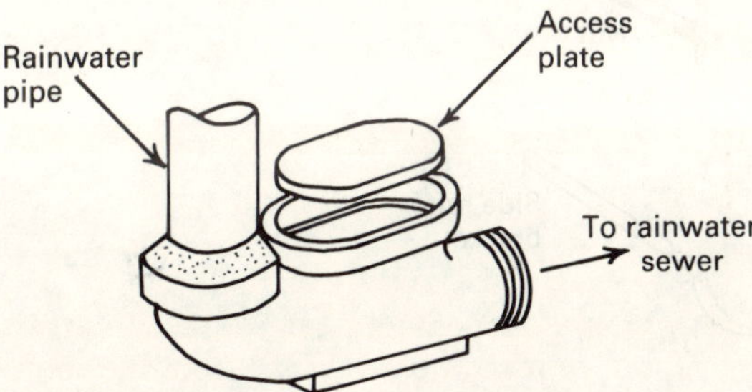

Fig. 4.2 Rainwater shoe in clayware

purpose'. There are numerous methods for rainwater gutter and downpipe design, most producing comparable results. Some are no more than a crude rule of thumb, others graphical or mathematical. For further information on roof drainage calculations see BS CP 308 : 1974.

Section (b) states that the gutter and rainwater pipe are to be 'composed of suitable materials of adequate strength and durability'. The most common material in recent years is cast iron, but this has now been superseded by uPVC. Rainwater pipes are either round or square, but gutter designs vary. Some examples of gutter profile are shown in Fig. 4.3.

Section (c) states that the gutter and rainwater pipe must be 'adequately supported throughout its length without restraining thermal movement, . . . such support being securely attached to the building'. Certain gutters are secured directly to the fascia board or supported by projecting corbelled brickwork; the majority, however, require bracketing and examples are shown in Fig. 4.4. Spacing of

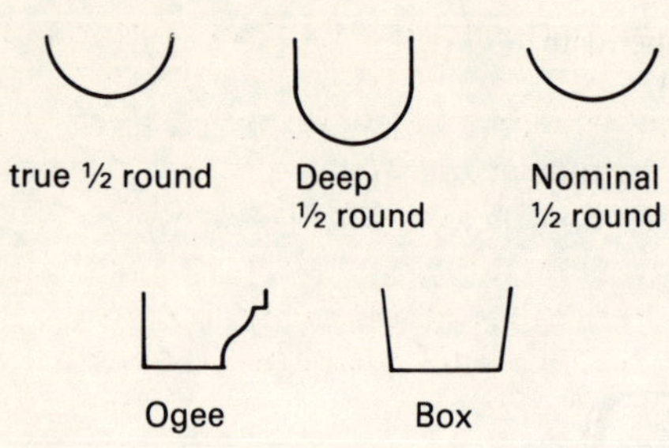

Fig. 4.3 Gutter profiles

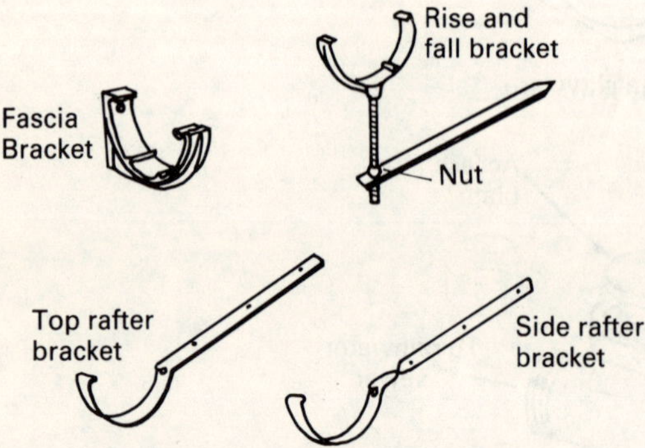

Fig. 4.4 Gutter brackets

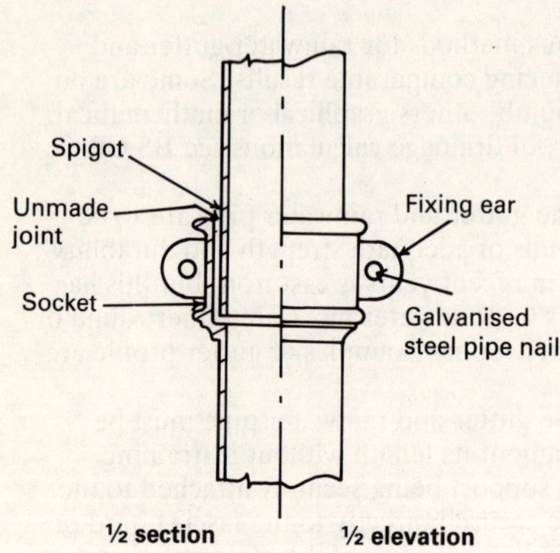

Fig. 4.5 Cast-iron rainwater pipe, socket with ears

brackets should not exceed 900 mm, particularly with uPVC guttering, where 600 mm is preferred.

Cast-iron down pipes are often provided with ears cast on the socket (see Fig. 4.5). These pipes are produced in 0.9, 1.2 and 1.8 m lengths, and support at socket alone is sufficient. Plastic pipe is produced in 2 m lengths, and bracketing of the type illustrated in Fig. 4.6 is advisable in the centre of the pipe length in addition to socket fixing. Thermal movement is a characteristic of uPVC systems,

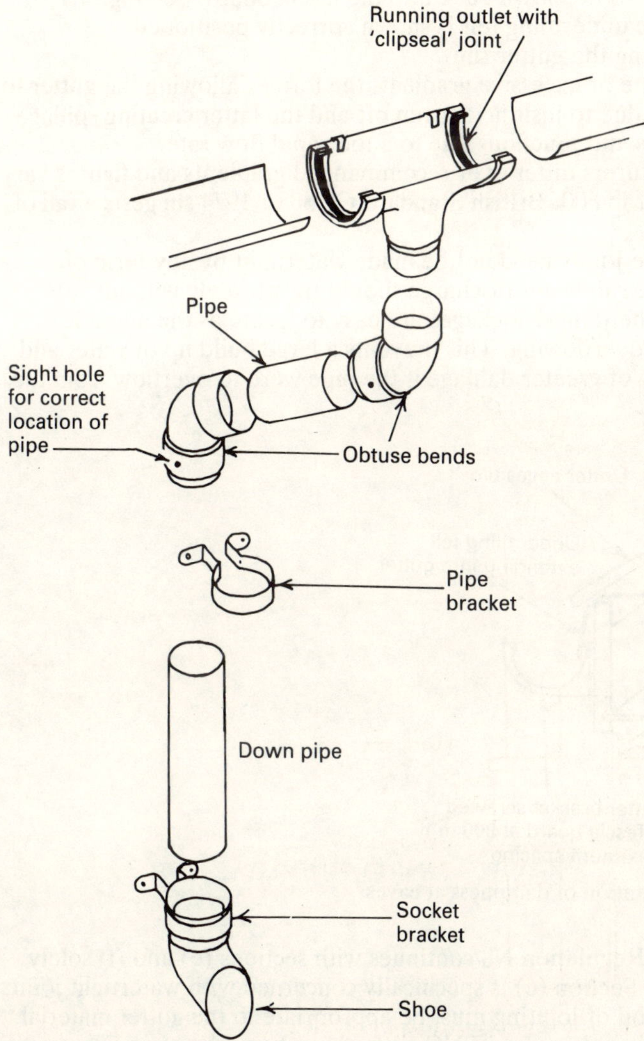

Fig. 4.6 'Osma-roundline' gutter and down pipe system

and the simple clip joint shown in Fig. 4.6 incorporates a synthetic rubber sealing pad which permits thermal movement whilst still retaining a watertight joint. Most plastic rainwater drainage systems require a 10 mm expansion gap at all joints including the push-fit spigot/socket downpipe joints shown in Fig. 4.6.

Section (d) requires the gutter and downpipe to be, 'so arranged as not to cause dampness in, or damage to, any part of the building'. With guttering, dampness may be caused by:

(a) The gap between eaves tiling and gutter being too large, causing rainwater to be blown back into the fascia board. See Fig. 4.7, where the undertiling felt is shown correctly positioned overlapping the gutter rim.

(b) Inadequate or excessive gradient, the former allowing the gutter to overflow due to insufficient run off and the latter creating spillage at corners and junctions due to a too rapid flow rate. Manufacturers differ over recommended gradients and figures vary down to 1 in 600. British Standard CP 308 : 1974 suggests a fall of 1 in 360.

Downpipe joints need not be made watertight by any form of sealant, as the rainwater discharge should travel freely without pressurising the joint. Blockages are easy to locate as the unmade joints permit overflowing. This prevents a large build up of water and the possibility of greater damage if the pipe were to overflow from the gutter outlet.

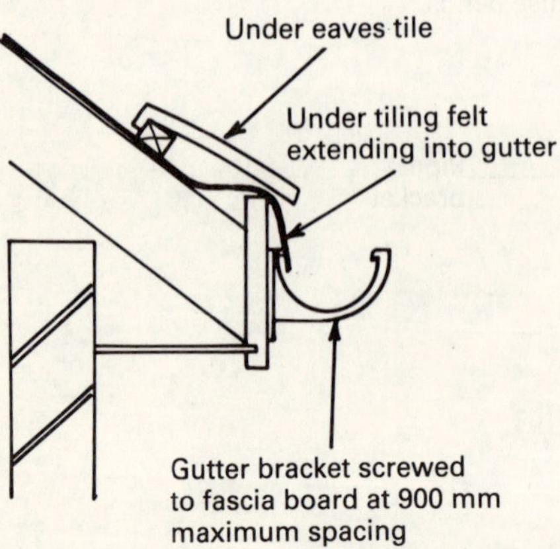

Fig. 4.7 Prevention of dampness at eaves

Building Regulation N8 continues with sections (e) and (f) solely for guttering. Section (e) is specifically concerned with watertight joints and the method of jointing must be appropriate to the gutter material. Synthetic rubber strips for uPVC gutter have been shown in Fig. 4.6. Metallic gutter systems require a bituminous mastic or red lead putty sealant, secured with non-ferrous or plated countersunk nut and bolt between the half round spigot and socket.

Section (f) requires sufficient number of outlets, which must be arranged to drain the complete gutter system. This may be established from design data, but position may be determined by aesthetic considerations or convenient location to the drainage system.

Sanitation

It is essential that any sanitary appliance connected to a drain is fitted with a water-filled trap. A trap is designed to prevent the entry of foul air from the sewer into the building. Some traps are an integral part of an appliance, a good example is the 50 mm water seal incorporated into a washdown WC pan. Other appliances have traps attached to them, they are situated underneath where access for maintenance will be necessary. Traps also have the function of retaining materials, which could otherwise obstruct the water pipe. Access to clear trap blockage is shown in Fig. 4.8, which indicates both tubular and bottle trap patterns.

In the interests of public health it is essential that the water seal is retained. Building Regulation N3 insists that the drainage system is designed 'to prevent the destruction under working conditions of the water seal in any trap in the system'. However, seal loss may occur due to poor design and workmanship. Possible causes of seal loss include self siphonage, induced siphonage and back pressure. See Fig. 4.9 for a diagrammatic explanation of these terms.

Traditional methods of seal retention include the installation of anti-siphon or vent pipes. These branch from the soil or waste discharge pipe in the manner shown in Fig. 4.10. When anti-siphon pipes are attached to every sanitary fitting, the resulting system is very

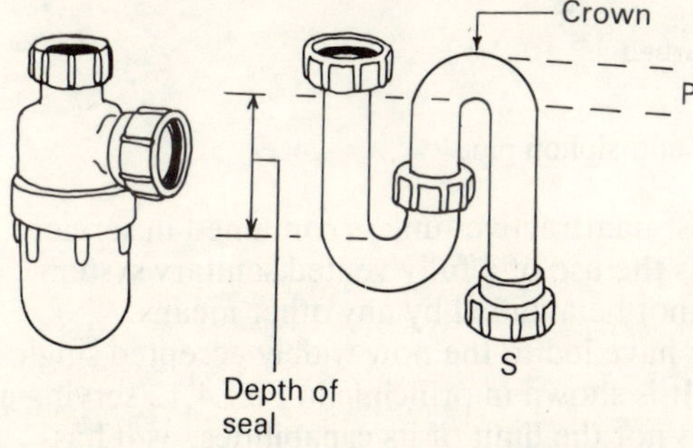

Fig. 4.8 Traps

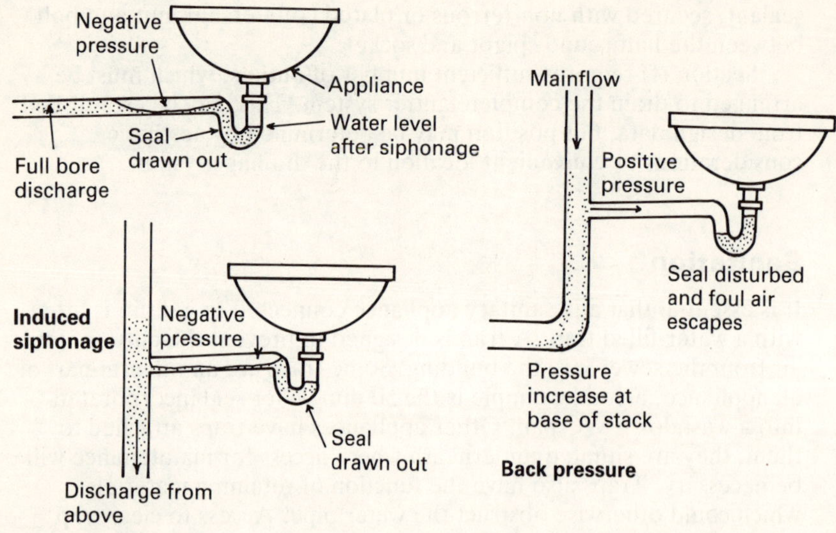

Fig. 4.9   Loss of seal in traps

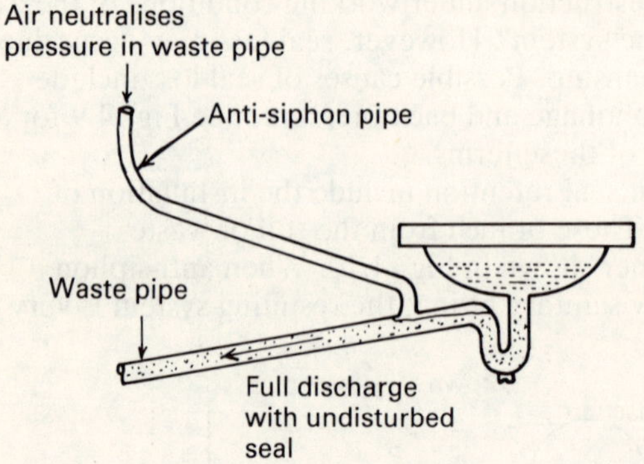

Fig. 4.10   Seal retention with anti-siphon pipe

expensive to install and most unattractive, unless contained in service ducts. Figure 4.11 illustrates the use of a fully vented sanitary system suitable where seal loss cannot be avoided by any other means.

Many years of research have led to the now widely accepted single stack system of sanitation. It is shown in principle in Fig. 4.12 serving a two-storey house, but this is not the limit of its capabilities, as it has potential for use in flats, offices and industrial buildings. The system shown incorporates the design recommendations of BS 5572 : 1978, which emphasises the features of low gradient branch pipes with limited lengths.

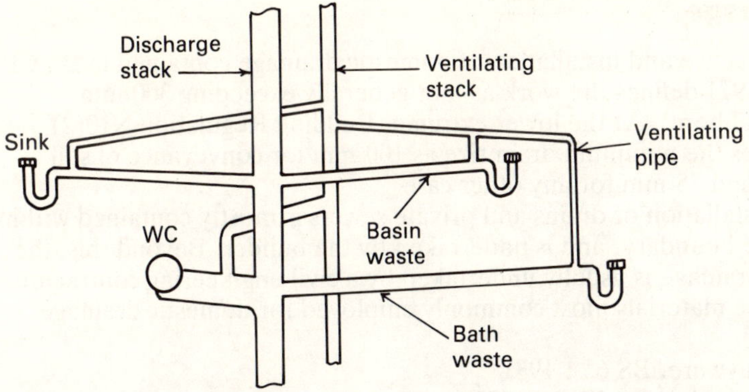

Fig. 4.11 Fully vented system of sanitation

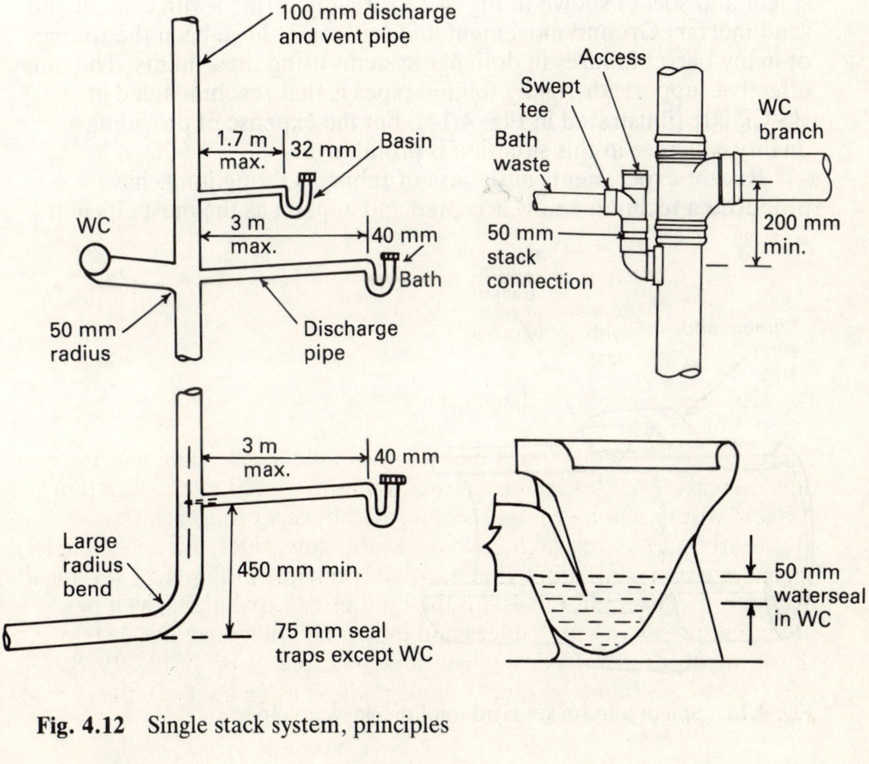

Fig. 4.12 Single stack system, principles

Drainage

The practice and installation of domestic drainage contained in BS CP 301 : 1971 defines the work as 'not generally exceeding 300 mm nominal bore'. At the lower extreme, Building Regulation N10(2) specifies the minimum drain size as 100 mm for conveyance of soil water and 75 mm for any other case.

Installation of drains and private sewers is mostly contained within the site boundary, and is undertaken by the builder. Beyond this, the main drainage is usually undertaken by a civil engineering contractor.

The materials most commonly employed for domestic drainage include:

Clayware, BS 65 : 1981
Pitch fibre BS 2760 : 1973
Unplasticised polyvinyl chloride, BS 4660 : 1973.

Clayware

The British Standard prefers sizes up to 300 mm. Nominal bore are 100, 150, 225 and 300 mm. Clayware pipes have a long history associated with the conveyance of fluids. As a modern means of drainage they are used with spigot and socket joints or in plain end form with push-fit polypropylene sleeves. The standard joint between spigot and socket shown in Fig. 4.13 is rigidly formed with cement and sand mortar. Ground movement and settlement have been the source of many barrel failures in drainage systems using these joints. The only effective support for rigidly jointed pipes is that recommended in BS CP 301 (illustrated in Fig. 4.14), but the expense of providing quality concrete in this situation is prohibitive.

Recent experiments in the use of rubber 'O' ring joints have produced a technique now accepted and applied as the most efficient

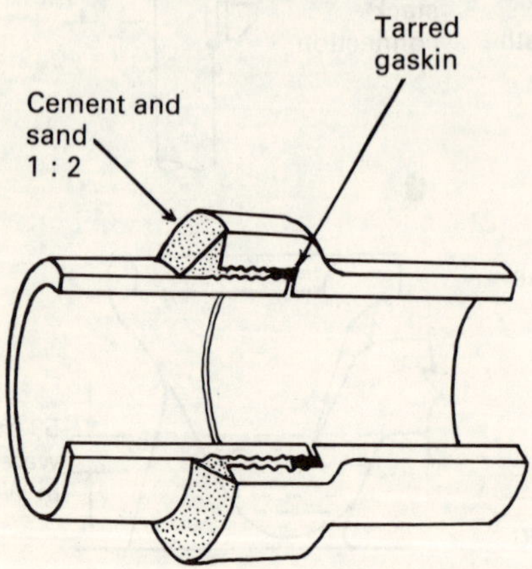

Fig. 4.13   Spigot and socket rigid joint on clayware pipe

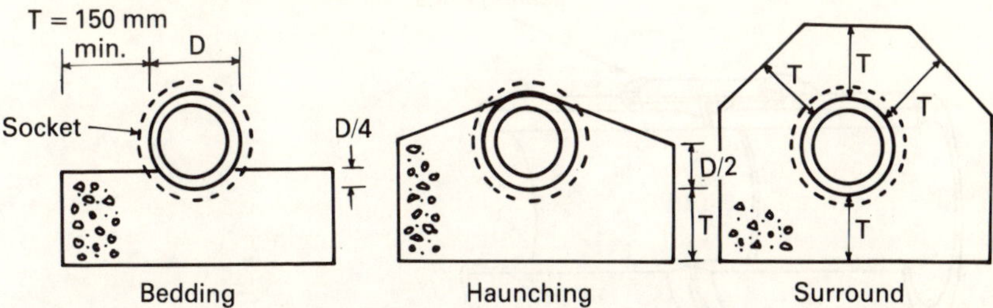

Fig. 4.14 Concrete support and protection to drainpipes

and economic means of jointing clay pipes. The 'Hepseal' joint, shown in Fig. 4.15, uses the principle of spigot and socket jointing, but the inside of the socket and rim of the spigot have polyester mouldings designed to locate a rubber 'O' ring gasket. This provides a watertight joint with sufficient angular movement and linear draw for all but extreme conditions. The nominal bore sizes up to 300 mm are 150, 225 and 300 mm in effective pipe lengths of 1.5 m.

An alternative form of flexible joint uses the 'Hepsleve' system shown in Fig. 4.16. Plain ended pipes are jointed with polypropylene flexible couplings containing sealing rings to locate and make watertight adjacent sections of pipe. Pipes using this technique are available in 100 and 150 mm, internal diameters, and in standard effective lengths of 1.6 m.

Pitch fibre

These are of BS sizes 50, 75, 100, 125, 150, 200 and 225 mm internal diameter. 50 mm is not suitable for drainage, (Building Regulation

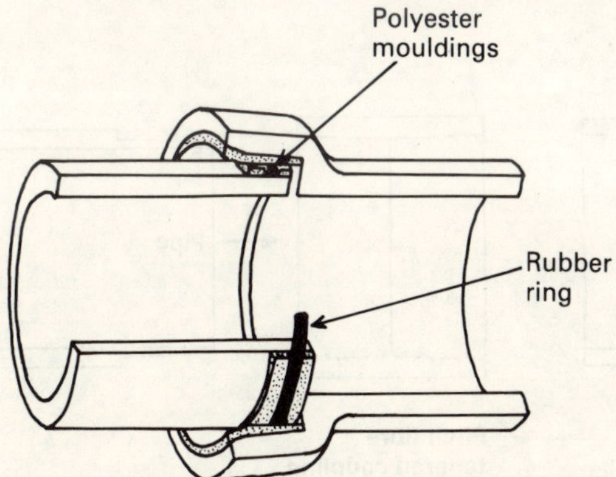

Fig. 4.15 'Hepseal' flexible joint on clayware drainpipe

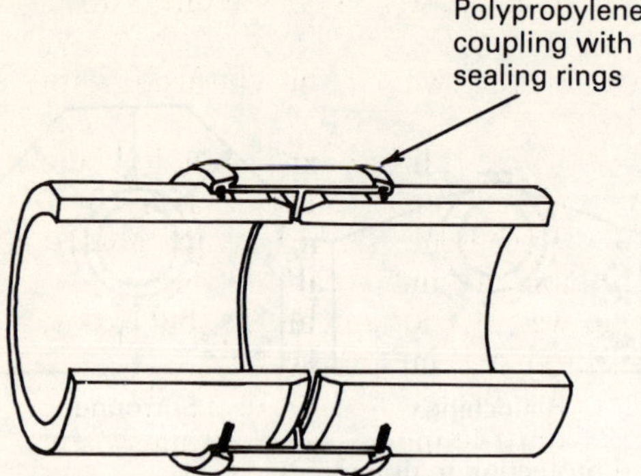

Fig. 4.16 'Hepsleve' flexible joint for plain end clay pipes

N10(2)). This size is manufactured primarily for use in conduit systems.

Pitch fibre pipes have a much shorter history than clayware, being first commercially produced in this country in the early 1950s. They are made from wood cellulose fibre, vacuum impregnated with coal tar pitch or bitumen. Properties include resistance to chemicals, corrosive soils, frost, vermin and the ability of the pipe as well as the joint to absorb ground movement without rupture. The standard overall length is 2.45 m for all sizes except 100 mm where a 3.05 m length is also possible. Jointing is by push-fit couplings of the snap joint or tapered joint type illustrated in Fig. 4.17.

Unplasticised polyvinyl chloride (uPVC)

British Standard sizes of uPVC are 110 and 160 mm external diameter. Other non-standard sizes include, 200, 250 and 300 mm nominal bore.

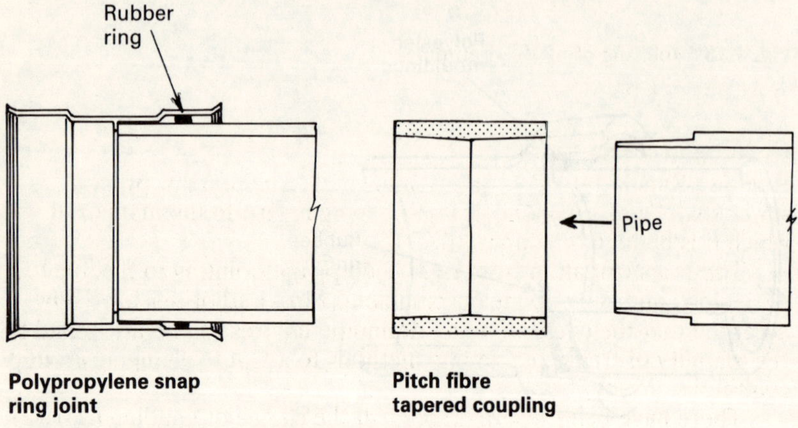

Fig. 4.17 Jointing of pitch fibre drainpipes

Lengths are specified as 1, 3 and 6 m, but some manufacturers also produce 2 and 9 m lengths.

The colour is a distinctive golden brown to avoid confusion with the light grey sanitary pipework.

This material has a high coefficient of linear expansion, but below the surface this is only a problem if high-temperature fluids are involved. Distortion may also be a problem with high temperatures; it may therefore not always be suitable, for industrial use. It is impervious to normal household waste, acids and alkalis, but exposure to abrasive materials may deteriorate the internal surface.

Jointing is by solvent weld or the preferred push-fit 'O' ring coupling as shown in Fig. 4.18. Constructing the whole drain system with welded joints is unacceptable, as this restricts thermal expansion and contraction. Solvent weld joints should only occur when assembling or making up fittings. The pipe itself is sufficiently flexible to absorb ground movement, but under excessive load will deform. Hence a depth of 1.2 m below surface level is recommended unless surrounded with concrete – in other situations a bed, side fills and blanket of suitable granular material will be required.

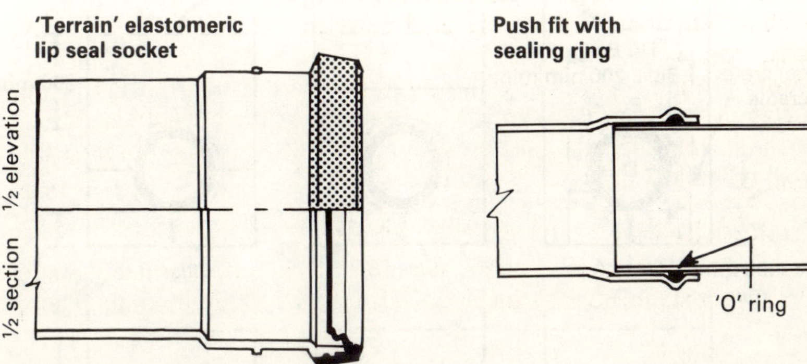

Fig. 4.18   Jointing of uPVC drainpipes

Bedding of drain pipes

Building Regulation N10 (1) (a) states that 'Any drain or private sewer shall be of sufficient strength having regard to the manner in which it is bedded or supported'.

The transformation from rigid bedding and jointing to flexible has been slow, and is by no means complete. Most authorities have now accepted that the cause of many drainpipe failures can be attributed to the inability of traditional laying methods to adapt to changing ground conditions.

There have been several papers on the choice of bedding for rigid pipes with flexible joints and a comparison of the suggested bedding

techniques, classes A, S, B, Fss and D contained in recent
governmental publications are shown in the table below and
Fig. 4.19. Some are less practical than others, but notice how the
bedding factors vary:

Class		Bedding factor
A	– Reinforced concrete cradle*	3.4
A	– Plain concrete cradle*	2.6
S	– 360° surround granular material	2.2
B	– 180° support granular material	1.9
Fss	– Flat layer single size granular material	1.5
N	– Flat layer all in aggregate or other suitable material	1.1
D	– Natural trench bottom	1.1

* Approximately 12 mm wide vertical gaps should be formed in the concrete at
pipe joint positions as spacings dependant upon the amount of ground
movement likely to occur but, in any case, not greater than 5 m.

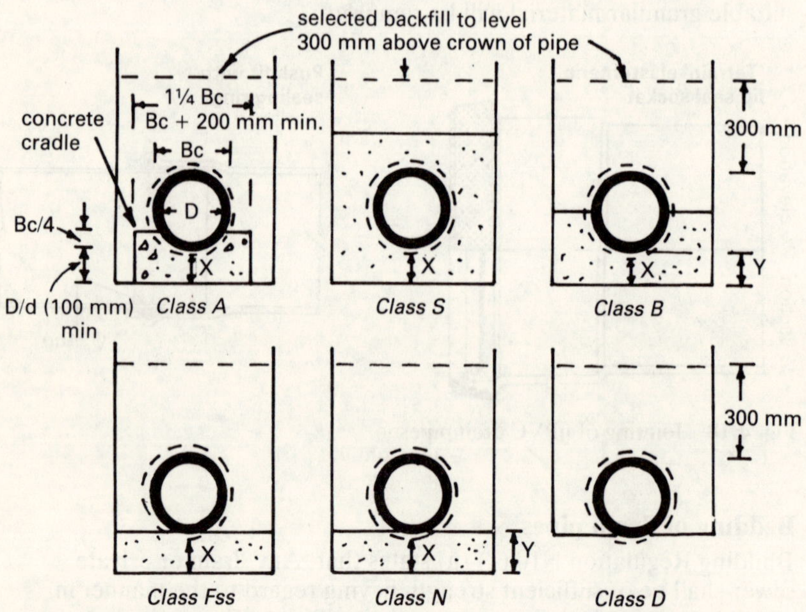

Y = 100 mm min. or Bc/6 in uniform soil (50 mm for plain ended pipes with
sleeve couplings)
= 200 mm min. or Bc/4 in rock or mixed soils containing rock etc. (150 mm for
plain ended pipes with sleeve couplings)
X = 50 mm min. under sockets in uniform soil
= 150 mm min. under sockets in rock or mixed soils containing rock etc.

Fig. 4.19 Techniques of bedding rigid pipes with flexible joints

The bedding factor is the ratio of vertical loading the pipe will carry, to the test load specified in the British Standard and a suitable combination of bedding and pipe strength is selected to resist the effects of backfill and surcharge load i.e. Class B is 1.9 times better.

Provision for access to drains

Building Regulation N12 (1) states that 'Any drain or private sewer shall have such means of access as may be necessary for inspection and cleansing'. This regulation continues by specifying the mandatory position of inspection chambers on drains and private sewers.
These are:

1. Changes in direction.
2. Changes in gradient.
3. 12.5 m maximum, from the junction with another drain or sewer, unless an inspection chamber is provided at that junction.
4. At the highest practicable point on a private sewer, unless there is a rodding eye (see Fig. 4.20) situated here.
5. At a maximum of 90 m apart.

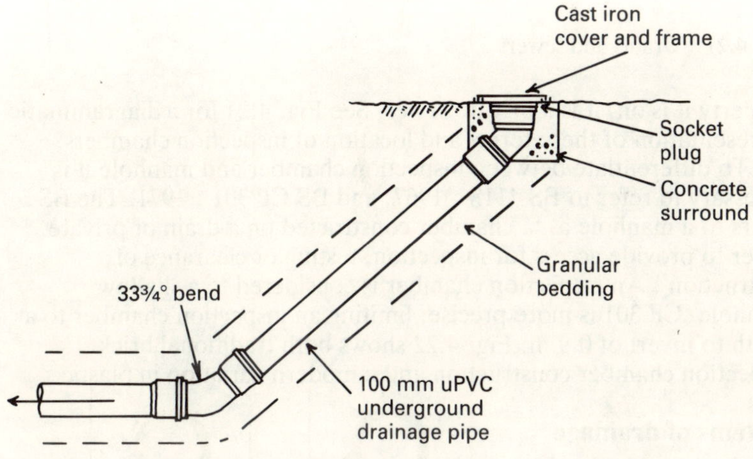

Fig. 4.20   Rodding eye access at head of drain

It is useful to distinguish between drain and private sewer, but this is an area where positive definitions are difficult as similarity in drainage systems is unusual. To generalise, they are both drainage branches, the private sewer serving more than one area of ownership before converging on the local authority main sewer. A drain is the branch off a private sewer to individual dwellings or a group of buildings within the same curtilage. Thus, a drain only remains so until it combines with the discharge from another drain. At that point, regardless of whose

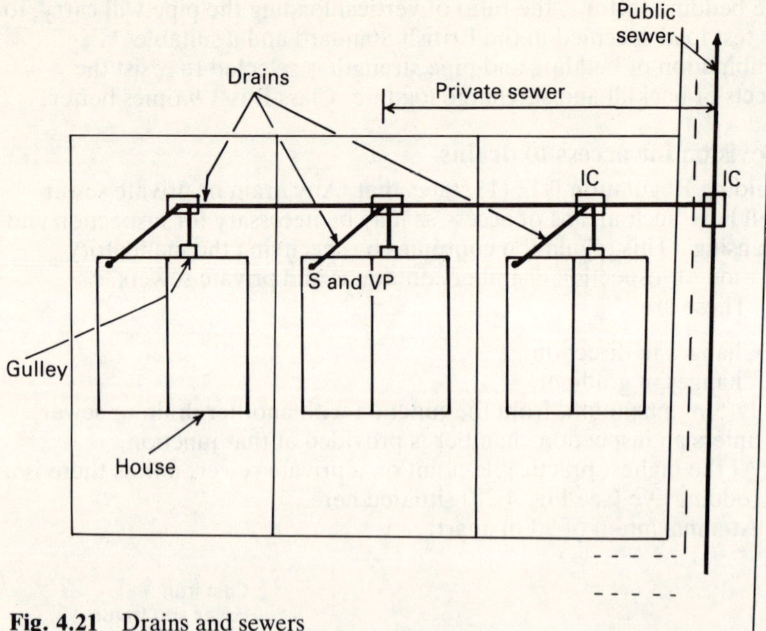

Fig. 4.21 Drains and sewers

property it is on, it is a private sewer. See Fig. 4.21 for a diagrammatic
representation of these terms and location of inspection chambers.

To differentiate between inspection chamber and manhole it is
necessary to refer to BS 4118 : 1967, and BS CP 301 : 1971. The BS
refers to a manhole as 'a chamber constructed on a drain or private
sewer to provide access for inspection, testing or clearance of
obstruction'. An inspection chamber is considered to a shallow
manhole. CP 301 is more precise, limiting an inspection chamber to a
depth to invert of 0.9 m. Fig. 4.22 shows both traditional brick
inspection chamber construction and a modern variation in plastic.

Systems of drainage

Two basic methods of foulwater and surface water conveyance exist,
the combined and separate systems. The combined system has a
common drain conveying both foulwater and surface water to a
combined sewer. This is a much more economical means of drainage
than the separate system, where a foulwater drain and a surface water
drain are necessary for separate conveyance to a foulwater sewer and a
surface water sewer. The combined system benefits further, by
thorough flushing during periods of rainfall, but a larger diameter pipe
may be necessary for the additional volume. Very few local authorities
will accept the combined system, as the increased volume due to
surface water will incur unnecessary treatment costs. Also, their
sewage treatment plant may not have sufficient capacity.

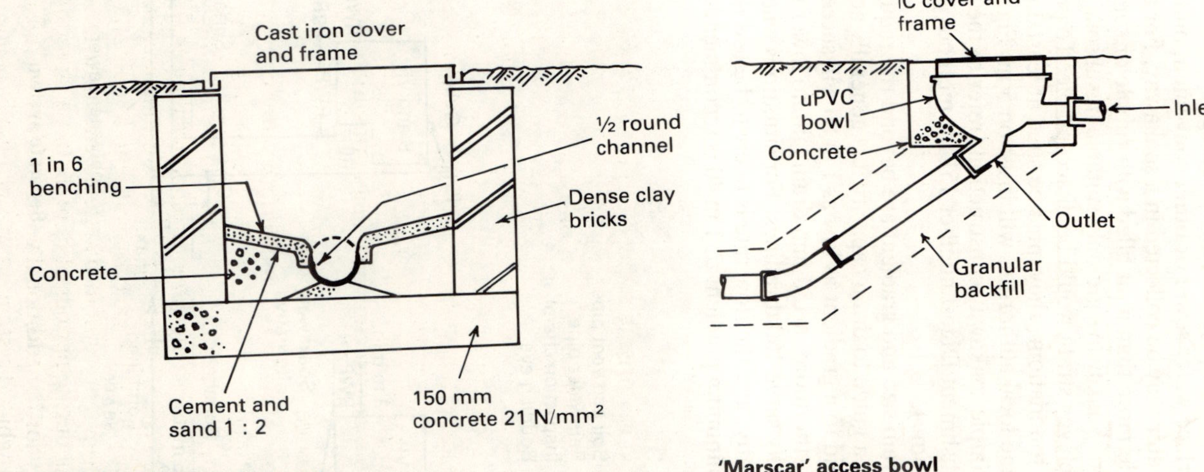

Traditional brick and concrete

'Marscar' access bowl

Fig. 4.22 Inspection chambers

A partially separate system is occasionally acceptable, where in the interests of convenience one rainwater pipe is permitted connection to a foulwater drain. See Fig. 4.23 for a comparison of both combined and separate systems serving the same dwelling.

Where a surface water sewer or other convenient outfall is not available, rainwater may be collected in a soakaway. For a dwelling house this is little more than a pit filled with rubble or some other easy draining medium. Soakaways are only suitable in subsoils that drain well, therefore clayey strata would be unacceptable. To prevent undermining of foundations, they must be sited at least 3 m from the building and some local authorities will insist on 5 m. Fig. 4.24 shows an example of simple soakaway construction (refer to the Building Research Establishment Digest number 151 for methods of determining volume).

Design of drain size and gradient are beyond the requirements of this level, but as a guide to drain capacity it is generally accepted that a 100 mm drain laid to a gradient as low as 1 in 80 will successfully serve 20 dwellings. Furthermore, a 150 mm drain laid no lower than 1 in 150 should successfully serve 100 dwellings. For situations where the drain is used infrequently, and the flow rate is likely to be less than 1 litre/second, a minimum gradient of 1 in 40 is recommended.

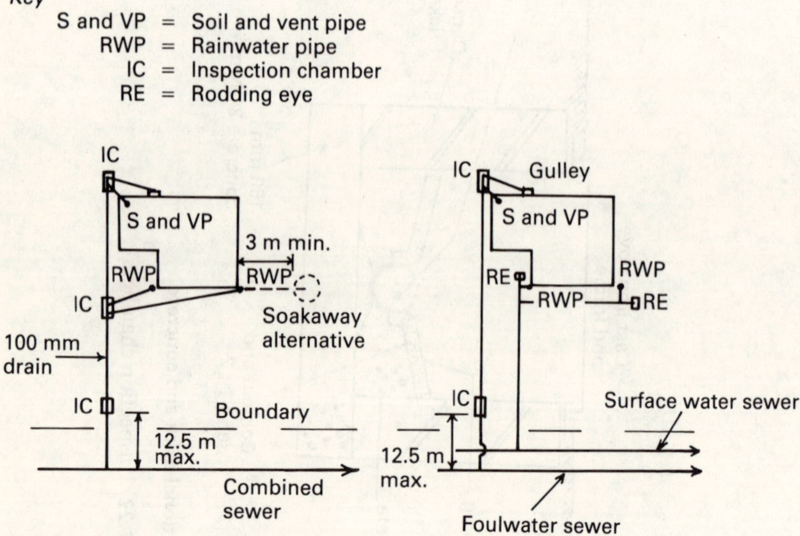

Fig. 4.23 Systems of drainage

D = depth below invert level,
approximately equal to d, the diameter.

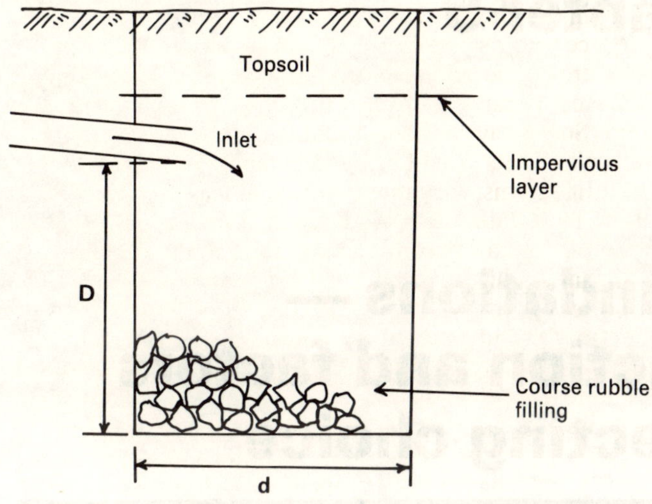

Fig. 4.24 Simple filled soakaway

Chapter 5

Foundations — function and factors affecting choice

Primary functions of a foundation

The fundamental function of the foundation to a building is to provide
a stable base and uniform support.

It is compulsory for the builder to satisfy these requirements and
this is defined in Building Regulation D3, para. (a). 'The foundation of
a building shall . . . safely sustain and transmit to the ground the
combined dead load, imposed load and wind load in such a manner as
not to cause any settlement or other movement which would impair the
stability of, or cause damage to, the whole or any part of the building
or of any adjoining building or works.'

Loading factors are determined in Building Regulation D.1. 'Dead
load means the force due to the static mass of all walls, partitions,
floors, roofs and finishes, including all other permanent construction.'
This, therefore, refers to the entire weight (converted to a unit of
force) of the structure as built, and takes no account of the contents or
forces directed by extremes of weather. British Standard Code of
Practice CP3 : Ch. V : Pt 1 : 1967 states that 'Dead loads shall be
calculated from the unit weight . . . or from the actual known weights of
materials used.'

'Imposed load means the load assumed to be produced by the
intended occupancy or use, including distributed, concentrated,
impact, inertia and snow loads, but excluding wind loads.' For dwelling
house construction this covers forces exerted by furniture and fixtures
in addition to loading by persons using the building. This is determined
in CP3 : Ch. V : Pt 1 : 1967 as a distributed floor load of 1.5 kN/m^2

intensity, and where a concentrated load is anticipated, 1.4 kN over a 300 mm sided square. Alternatively, the Building Regulations permit a distributed load of 1.44 kN/m^2 provided the house is for one family use only and is contained within three storeys.

Impact and inertia loading are design factors included in industrial and larger scale construction where dynamic loads from plant and machinery must be accounted for.

Snow loading varies according to the roof slope. In CP3 : Ch. V : Pt 1 : 1967 this varies between not less than 1.5 kN/m^2 for flat roofs and 0.75 kN/m^2 where the slope exceeds 30°.

Wind loads and their effect on a foundation are less easily defined. Positive and negative loads will vary around the structure, depending on the degree of wind strength, the wind direction and the shape of the building. Both Building Research Establishment Digest No. 119 and CP3 : Ch. V : Pt 2 : 1972 consider the assessment of wind load on

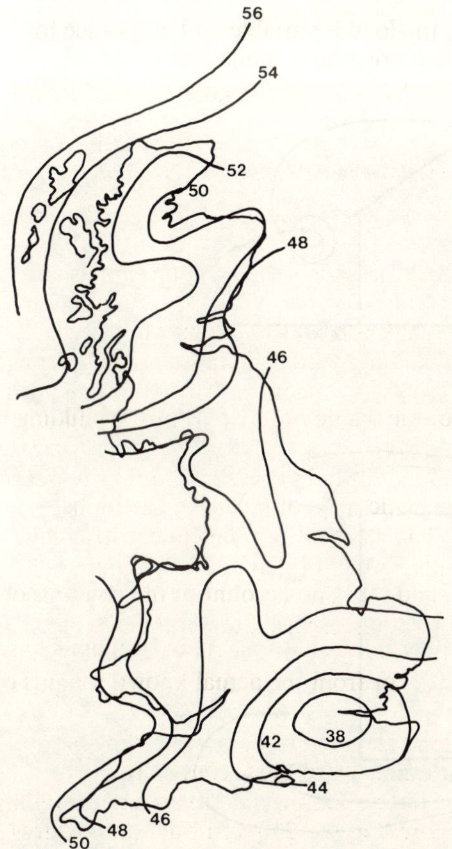

Fig. 5.1 Basic wind speed in metres per second

52

ordinary buildings and they base the resulting imposed load on the dynamic pressure of wind. Maximum wind speed (V) is determined from the isopleths shown in Fig. 5.1 on the map of the British Isles, and are corrected by factors for topography (S_1), ground roughness (S_2) and period of exposure (S_3). The latter factor is appropriate to temporary structures and circumstances of an exceptional nature. For normal buildings this is taken as unity and this is defined in para. (c) to Building Regulation D2 (2) as never less than 1.

Hence, design wind speed (Vs), is found from the formula:

$$Vs = V . S_1 . S_2 . S_3.$$

On walls exposed directly to wind, the dynamic pressure is calculated from the formula:

$$q = k . Vs^2 \qquad \begin{aligned} q &= \text{N/m}^2 \\ k &= \text{constant } 0.613 \\ Vs &= \text{m/s} \end{aligned}$$

However, as previously noted, the load is variable and allowance for variations is incorporated in the correction formula:

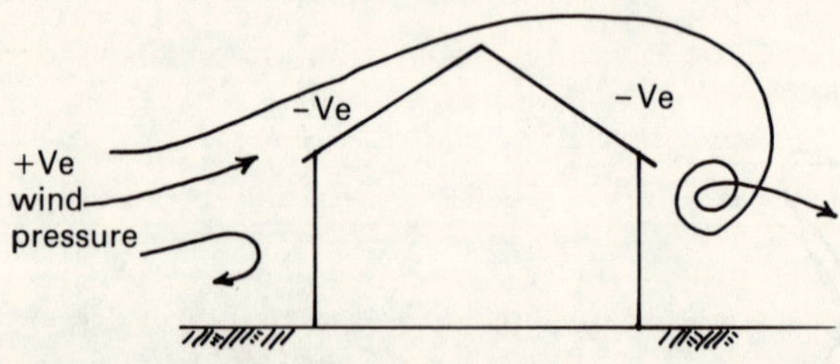

**Elevation with wind
square to building**

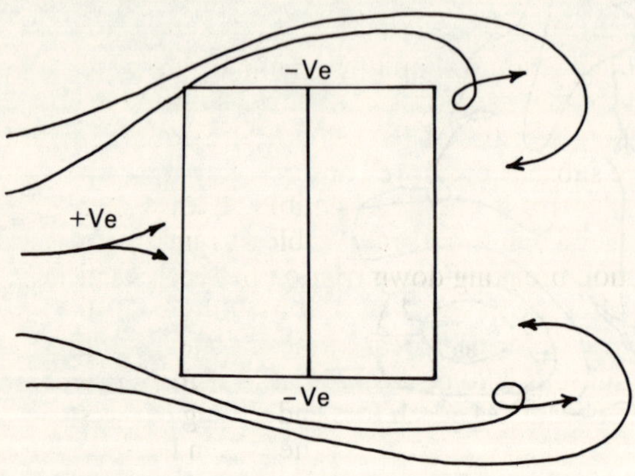

Plan

Fig. 5.2   Effect of wind load

$$p = Cp.q$$

p = pressure at the surface under consideration
Cp = pressure coefficient which may be negative.
(See section 7, CP3 and Fig. 5.2)

Choice of foundation

The foundation to a building is considered under two distinct headings:

1. The natural foundation subsoil which receives the building; and
2. The man-made or artificial foundation which bears the structural loads and transfers these to the subsoil.

There are many forms of artificial foundation, and the design and construction relates to:

(a) The composition of the natural foundation and its bearing properties
(b) The natural formation or level of the site
(c) The total load (dead, imposed and wind) which the foundation is expected to carry.

(a) For purposes of explanation and consideration, most subsoils are defined within either of two categories, according to whether their strength is due to cohesion or friction (non-cohesive). Cohesive soils contain minute particles of clay which are bound together with water. In most clays the strength and bearing capacity will be determined by the proportion of water. Non-cohesive soils contain larger particles and include sands, gravels and boulders, their strength obtained by particle weight and how well they are retained.
(b) Level sites of uniform subsoil will gradually absorb the superimposed loading of a new building. Unless there are unforeseen subsoil defects settlement will be the same throughout. The amount should be minimal, but in soft clays excessive settlement can occur in addition to swelling and shrinkage, characteristics of flat unnaturally draining subsoil. Non-cohesive subsoils drain well and resist compressive forces, therefore settlement is usually negligible on level sites. Problems occur if disturbed below the water table as sand may run due to the water action breaking down friction between particles.
    Sloping sites are never ideal, but building land shortage has released the use of formerly less attractive sites. Building on sloping clay subsoils is to be avoided, as disturbance will reduce the cohesion, causing strength loss and sliding of the structure. Sloping sandy sites also pose difficulties when loaded. Slip or shear failure caused by frictional breakdown between particles is not uncommon.

54

(c) The total load on a simple strip foundation is calculated by
     summating the loading and dividing this by the length of
     supporting foundation. However, not all structures are uniformly
     loaded and increases occur at attached piers, columns and
     chimneys. Furthermore, openings for windows and doors, gable
     ends, floor loads and roof loads will apply disproportional loading
     on different areas of the same foundation. It is impractical to
     consider varying loads for an individual building, so the worst
     situation is located and the maximum load per metre run of
     foundation calculated.

     As a guide to foundation width, the table to Building
     Regulation D7 provides a comparison between loading and subsoil
     quality. The subsoil is tested by the methods specified in column 3
     and the width of foundation located by comparison of results with
     loading. For precise determination of foundation width, subsoil
     analysis is necessary to confirm bearing capacity. This is related to
     wall loading in the following formula and width of foundation
     located by simple arithmetic. Figure 5.3 illustrates this method.

$$\text{Width of strip foundation} = \frac{\text{load per metre run}}{\text{safe bearing capacity of subsoil}}$$

e.g.

Bearing capacity of subsoil   = 100 kN/m^2
Factor of safety   = 2 kN
Loading per metre run of wall   = 25 kN

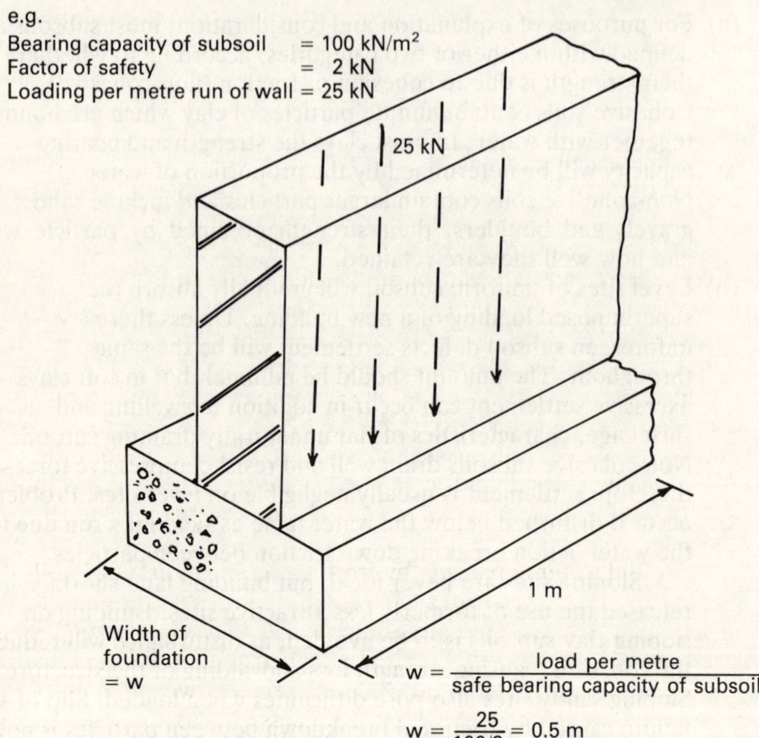

Fig. 5.3 Simple foundation design

Subsoil movement

Movement of the subsoil results from natural causes or inducement through excessive structural loading. Natural causes are numerous, and include the effects of shrinkage and swelling, trees and water. Structural loading will cause consolidation of the subsoil, and in some parts of the country subsidence due to the presence of former mine shafts is evident.

Shrinkage and swelling

All subsoils swell and shrink to some extent; the closer they are to the surface, the greater the likelihood of movement. This primary form of movement is directly related to the extent of dehydration, or, at the other extreme, frost. Building Regulation D3 para. (b) prevents foundation construction becoming affected by these extremes: 'the foundation of a building shall . . . be taken down to such a depth . . . to safeguard the building against swelling, shrinkage or freezing of the subsoil.'

The water content of many subsoils has a direct relationship with extent of movement. As water is removed, the particles become closer, an effect most noticeable in clays as these contain a large proportion of water. Reduction in water content increases the strength of clay, but creation of voids and cracks will seriously undermine foundations, as may be seen in Fig. 5.4. Conversely, the expansion (in many clays, over 25 per cent of the original volume) caused by water absorbtion reduces strength, and if irregular, imposes uneven forces on a foundation.

Volume changes are less noticeable in sands as the particles remain undiluted in the presence of water. The most significant effect that water has on sands is to wash away the finer particles, leaving a coarse sand of less stable structure. This is unlikely to occur on flat ground, but where water is draining from a higher level, typical of sloping sites, loss of ground by running sand renders a site useless for building. Ground water control techniques must be considered, if the site is to be viable for building purposes.

Furthermore, underground water sources can erode granular subsoils, creating voids into which the overburden collapses. A swallow hole forms above, causing serious undermining of the structure. Additionally, subsoil voids could be caused by an undetected leaking water main or by a rainwater soakaway situated too close to the building.

Ground water expansion by frost is possible in granular soils, particularly where the water table is high. However, even during the severest winters, frost penetration over 750 mm below surface level is unlikely, but the first few hundred millimetres are unprotected and may heave due to the effect of water freezing between the particles. Concrete foundations are unaffected unless too close to the surface and therefore contravening Building Regulation D3 (b). The problem area is with unheated buildings of raft or slab type construction, with granular filling to the underside.

Trees

Dehydration and shrinkage of the subsoil by the absorbing effect of trees effectively reduces the subsoil-bearing capacity. This is noticeable where the corners and sides of a building lean away slightly, causing descending cracks usually to weaker areas such as window and door openings. Movement of this form occurs because the inside of the building remains protected, whilst the perimeter wall and foundation are exposed. Cracks around foundations absorb water, dissolving the subsoil and reducing its bearing capacity.

Physical damage by growing tree roots is not uncommon; therefore, to avoid structural damage by either or both effects, location

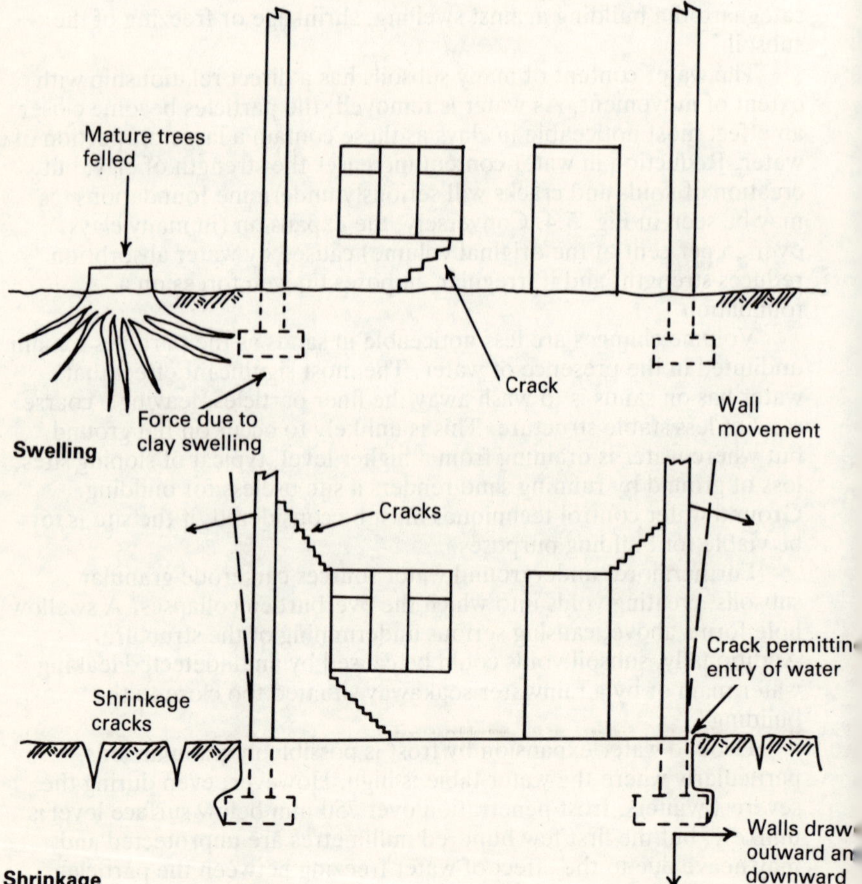

Fig. 5.4 Cracking associated with clay subsoils

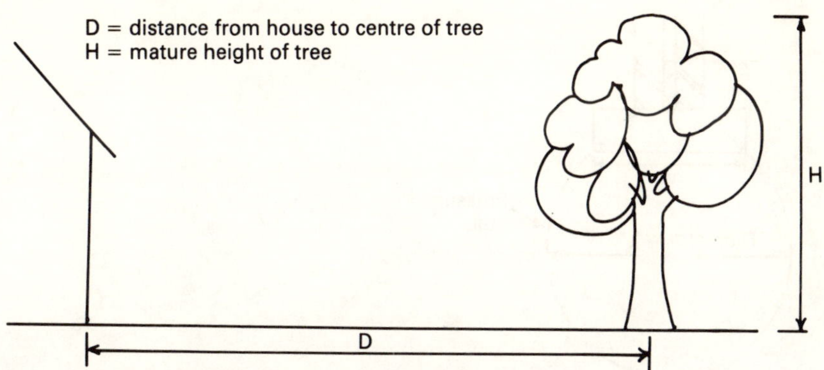

Species of tree	Distance from dwelling as proportion of tree height D/H							
	1/10	1/4	1/3	1/2	2/3	3/4	1	
Poplar, elm and willow	Not acceptable	2.8	2.6	2.3	2.1	1.9	1.5	Depth of foundation trench in metres
All others	Not acceptable	2.4	2.1	1.5	1.5	1.2	1.0	

Fig. 5.5 Proximity of trees: NHBC Rule Fo 19 (c)

of trees must be seriously considered. Building closer to a tree than its mature height is not recommended, and where several trees occur, the safe distance is accepted as one-and-a-half times the tree's mature height.

Swelling

Site clearance involving tree felling will have quite the opposite effect of dehydration. Mature trees extract enormous quantities of water from the ground, and large scale felling will seriously disturb the subsoil balance. Swelling will be the most obvious consequence, and construction should not commence until the ground is re-established. Clay soils subject to this form of imbalance will expand and exert excessive pressure on the underside of foundations. Where this is uneven, the effect is shown in Fig. 5.4, indicating differential swelling and cracking of parts of the structure.

Consolidation

Pressure on a subsoil from the building load will increase the water pressure beneath the foundation and force the water to percolate away until a balance is re-established. The subsoil particles compress and the building settles. This process is fairly rapid with sandy subsoils, but may continue for several years on clay. The area affected beneath a simple strip foundation is shown in Fig. 5.6 and is approximately one-and-a-half times the width of foundation.

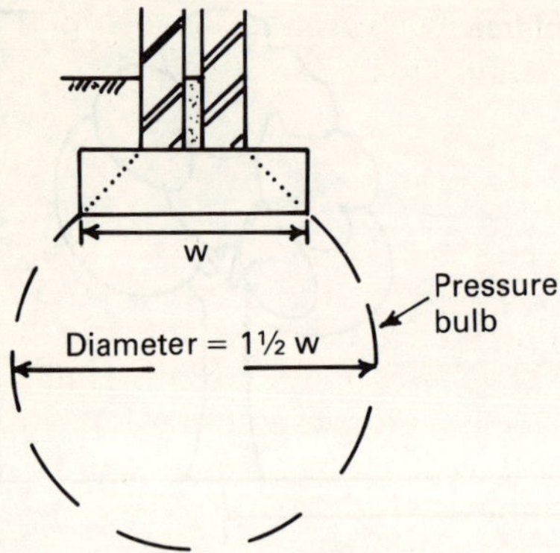

Fig. 5.6   Zone of subsoil affected by pressure on a simple strip foundation

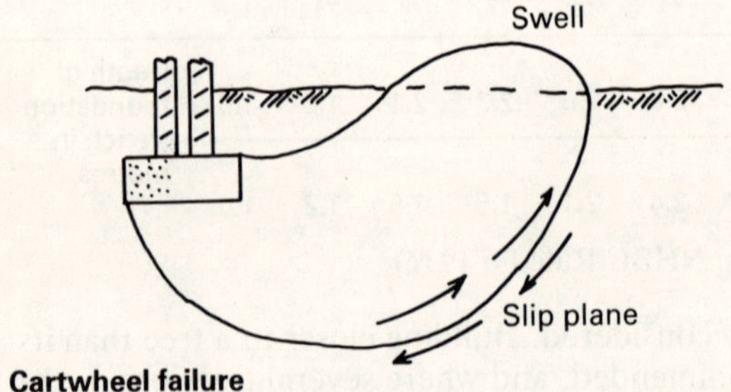

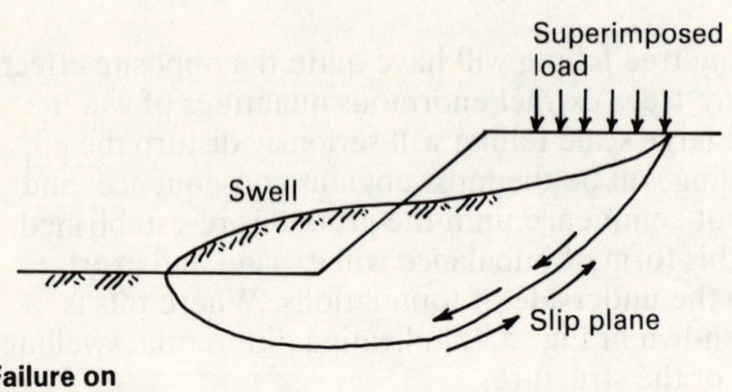

Fig. 5.7   Types of shear failure in subsoils

Shear failure of the subsoil

Failure occurs when the resistance to shearing stresses is exceeded and usually takes the form of a slip line. This frequently occurs where the subsoil absorbs excessive water and consequently softens.

Alternatively, a slip will occur where water penetration into cracks in clay subsoils follows long dry periods, or by over buoyancy of the subsoil near the surface due to a high water table. The effects are shown in Fig. 5.7.

Foundation materials

Modern artificial foundations are produced from concrete, with the addition of steel reinforcement where instability of the subsoil justifies it. Building Regulation D7 para. (c) specifies the concrete as 'composed of cement and fine and coarse aggregate conforming to BS 882 : 1965 in the proportion of 50 kg of cement to not more than 0.1 m^3 of fine aggregate and 0.2 m^3 of coarse aggregate.'

Foundation history reveals the use of many materials. Compacted

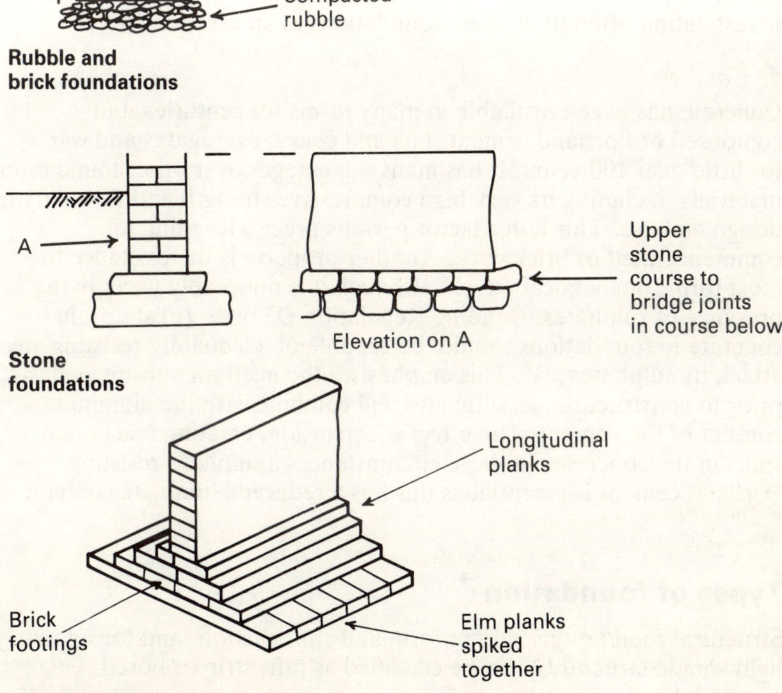

Fig. 5.8   Foundation history

rubble, stone slabs, bricks footings and timber planks illustrated in Fig. 5.8 are a few examples worthy of mention as many existing structures are supported by these.

Background to modern foundations

1. Rubble and brick foundations

The substructure contains little more than a shallow trench lined with well-compacted rubble to consolidate the subsoil. This is superimposed with bonded brick courses, gradually reducing in width to the finished wall thickness.

2. Stone footings

Usually two courses of stones, the upper course slightly shorter and laid to bridge the joints of the stones below.

3. Timber plank foundations

Often found under old buildings bearing on soft and wet subsoils. Invariably two layers of planks, the lowest layer square to the wall with upper layer parallel to and supporting the substructural masonry. Elm was always recognised for its virtual imperishable qualities in water and investigation often finds these foundations in an excellent state.

4. Concrete

Concrete has beeen available in many forms for centuries, but composed of Portland cement, fine and coarse aggregates and water, for little over 100 years. It has many advantages over other foundation materials, including its very high compressive strength and flexibility in design and use. This latter factor permits precise levelling for commencement of brickwork. Another property is its resistance to most forms of chemical attack, although it is noticeably weak in the presence of sulphates. Building Regulation D3 para. (c) states that concrete in foundations should 'be capable of adequately resisting any attack by sulphates . . .'. This emphasises the need for subsoil analysis prior to construction, as sulphates will combine with the aluminate content of the cement. The effect is expansion, creating cracks and splits in the concrete. In these circumstances a sulphate-resisting Portland cement is essential as this has a reduced aluminate content.

Types of foundation

Structural foundations vary in form and construction, and for relatively light simple structures may be classified as raft, strip or bored.

1. Raft foundations

British Standard Code of Practice, CP101 : 1972 defines a raft as 'A

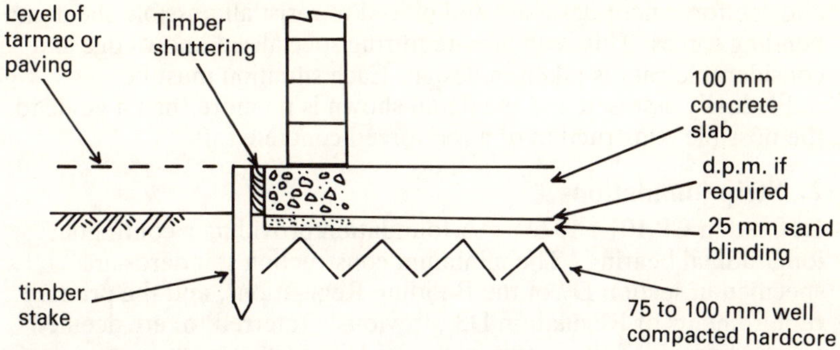

**Simple raft for light structures.
Edge section**

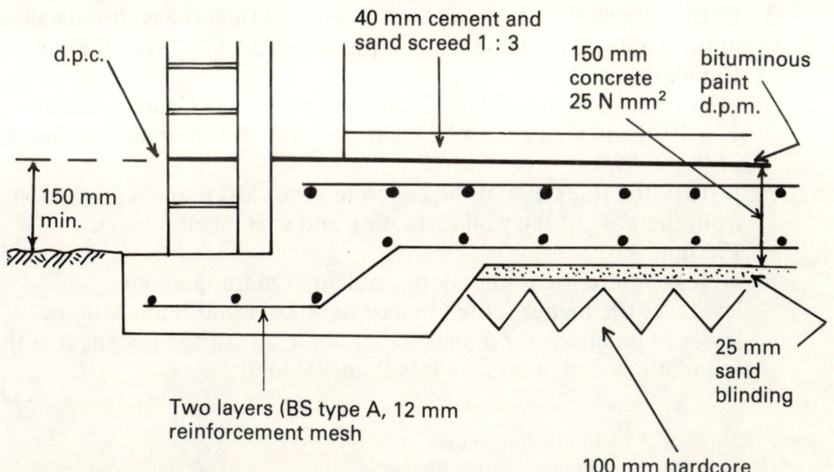

**Reinforced concrete raft foundation
for a small dwelling on poor subsoil**

Fig. 5.9 Raft foundations

foundation continuous in two directions usually covering an area equal to or greater than the base area of the structure.' In its simplest form this is little more than a concrete slab of about 100 mm thickness, constructed at ground level for uniform support of sheds, garages and similar lightweight buildings. Where the quality of subsoil is noticeably marshy or subject to the effects of minor subsidence, a steel-reinforced concrete raft designed to resist the tensile forces will provide a more substantial basis for construction than conventional strip. See Fig. 5.9 for examples of both simple and reinforced raft foundations.

The superstructure should be designed with loading concentric, and reinforcement designed and placed to resist all possible shear and bending forces. This is an area requiring specialised knowledge and considerable care is taken in design. Each situation must be individually assessed and the detail shown is no more than a guide to the possible construction of a reinforced concrete raft.

2. Strip foundations

Defined in CP 101 : 1972 as 'A foundation providing a continuous longitudinal bearing.' The minimum construction standards are specified in section D7 of the Building Regulations, and the primary requirements of Regulation D3, previously referred to, are deemed satisfied provided the requirements of D7 are observed:

1. 'the foundations . . . are . . . situated centrally under the walls,'
2. D7(a): 'there is no made ground or wide variation in the type of subsoil within the loaded area'
3. D7(b): 'the width of foundations is not less than the width specified in the Table'. This table may be found in Part D7 of the Building Regulations.
4. D7(c): see section headed *Foundation materials*. Cement to BS 12 : 1978 and water must be clean and free from impurities. See BS 3148 : 1959.
5. D7(d): 'the thickness of the concrete is not less than its projection from the base of the wall or footing and is in no case less than 150 mm'.
6. D7(e): 'where the foundations are laid at more than one level, . . . the higher foundations extend over and unite with the lower foundations for a distance of not less than the thickness of the foundations and in no case less than 300 mm'.

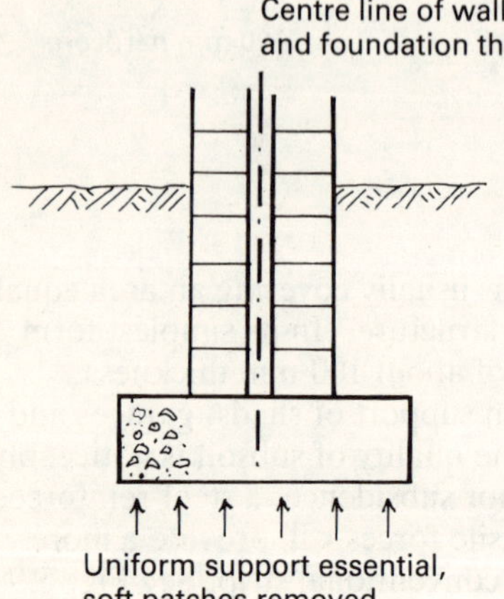

Fig. 5.10   Building Regulations D7 and D7 (a)

7. D7(f): 'where there is a pier, buttress or chimney forming part of a wall, the foundations project . . . on all sides to at least the same extent as they project beyond the wall'.

Figures 5.10–5.14 illustrate the effect and requirements of these regulations.

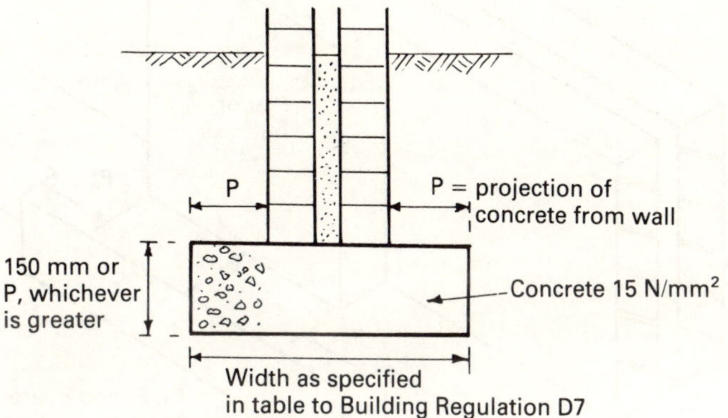

Fig. 5.11   Building Regulations D7 (b), (c) and (d)

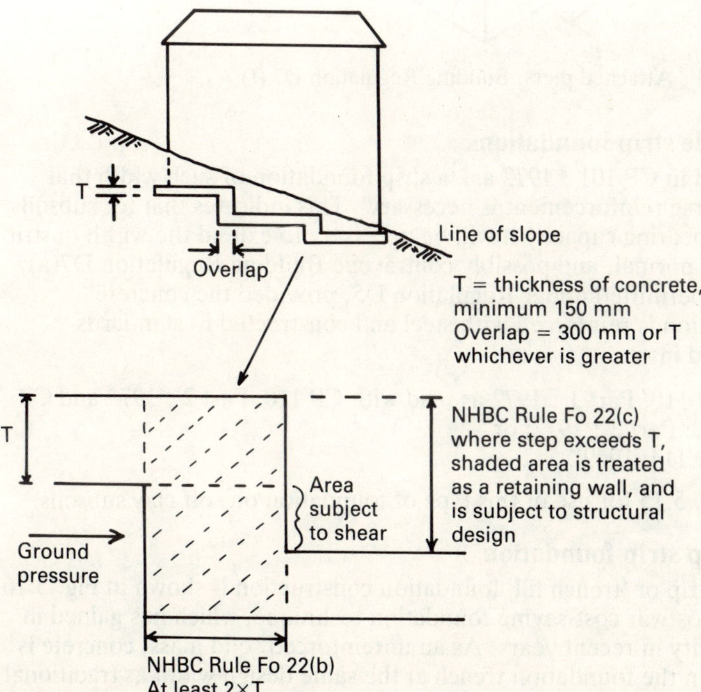

Fig. 5.12   Foundations on sloping sites. Building Regulation D7 (e)

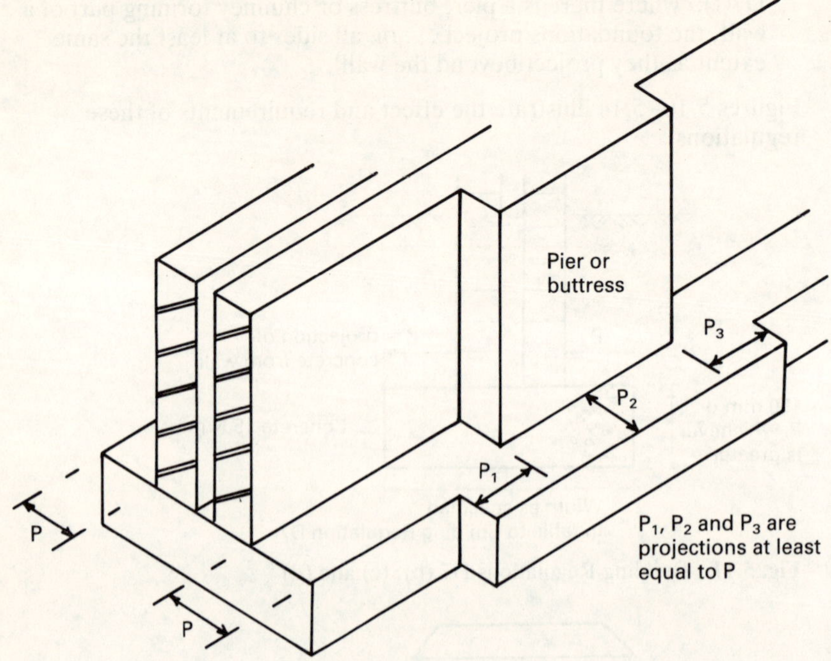

Fig. 5.13   Attached piers. Building Regulation D7 (f)

3. Wide strip foundations

Defined in CP 101 : 1972 as, 'a strip foundation of such width that transverse reinforcement is necessary'. This indicates that for subsoils of low bearing capacity it will be necessary to extend the width of strip beyond normal, and possibly contravene Building Regulation D7(a). This is permitted under Regulation D5, provided the concrete foundation is reinforced with steel and constructed to standards specified in:

'(a) CP 110: Part 1 : 1972 as read with CP 110: Part 2 : 1972 and CP 110: Part 3 : 1972; or
 (b) CP 114: 1969'.

See Fig. 5.15 for use of this type of foundation on soft clay subsoils.

4. Deep strip foundation

Deep strip or 'trench fill' foundation construction is shown in Fig. 5.16. It is a postwar cost-saving foundation technique, which has gained in popularity in recent years. As an unreinforced solid mass, concrete is placed in the foundation trench at the same design width as traditional strip, but with a thickness or depth up to reduced ground level. Considerably more concrete is used, but this is justified by the savings

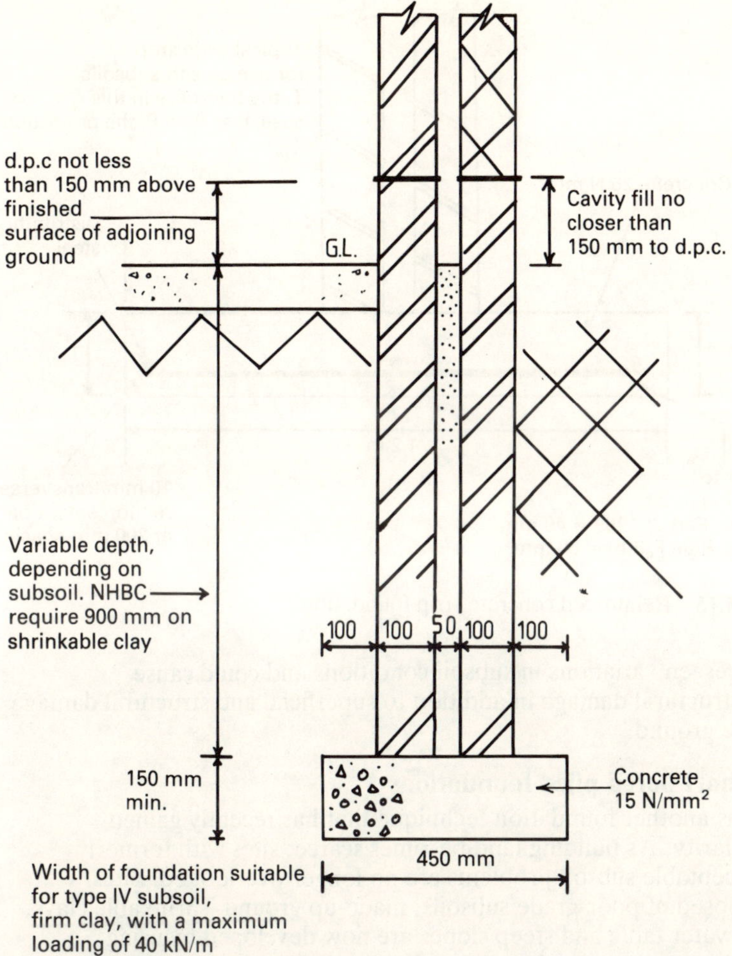

Fig. 5.14 Narrow strip foundation

from omission of trench support (assuming concrete is poured immediately following excavation), bricks and bricklaying time.

Alternatively, in poor subsoils, deep strip foundations are applied as a reinforced concrete beam, over areas subject to volume changes caused by seasonal differences. Reinforcement is not always required and excavation may be relatively narrow but deep, to maximise frictional resistance between foundation concrete and adjacent subsoil. The depth should be sufficient to isolate the foundation from surface swelling and shrinkage.

The use of reinforcement in poor subsoils is always advisable as assurance against differential structural settlement. This is caused by

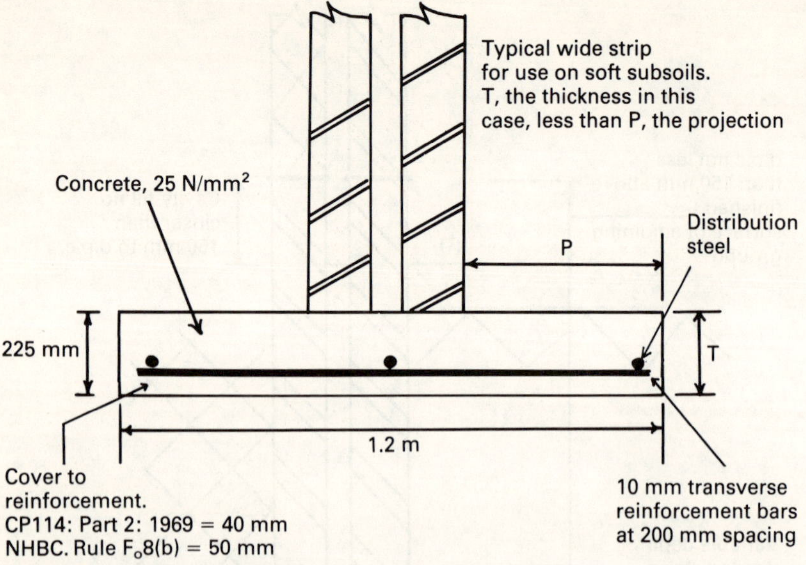

Fig. 5.15   Reinforced concrete strip foundation

unforeseen variations in subsoil conditions and could cause
sub-structural damage in addition to superficial and structural damage
above ground.

5. Short bored piled foundations

This is another foundation technique that has recently gained
popularity. As building land becomes scarce, sites with formerly
unacceptable subsoil problems are no longer overlooked. Sites
composed of poor-grade subsoils, made-up ground, shrinkable clays,
high water table and steep slopes are now developed by using
foundations supported on concrete piles bearing on strata below the
problem area. The principle is illustrated in Fig. 5.17. Construction
methods vary, and could include:

(a) Bored and cast *in situ* end-bearing concrete piles.
(b) Bored and under-reamed, cast *in situ* friction piles.
(c) Driven precast concrete end-bearing shell piles.

(a) Bored and cast in situ end bearing concrete piles

Boring is by lorry-mounted mechanical auger, or shell auger,
depending on site conditions and accessibility. Diameter of bore holes
varies between 250 and 350 mm, and length up to about 4 m. Steel
reinforcement is essential where ground movement is a possibility and
tops of piles are linked with reinforced ground beam to support

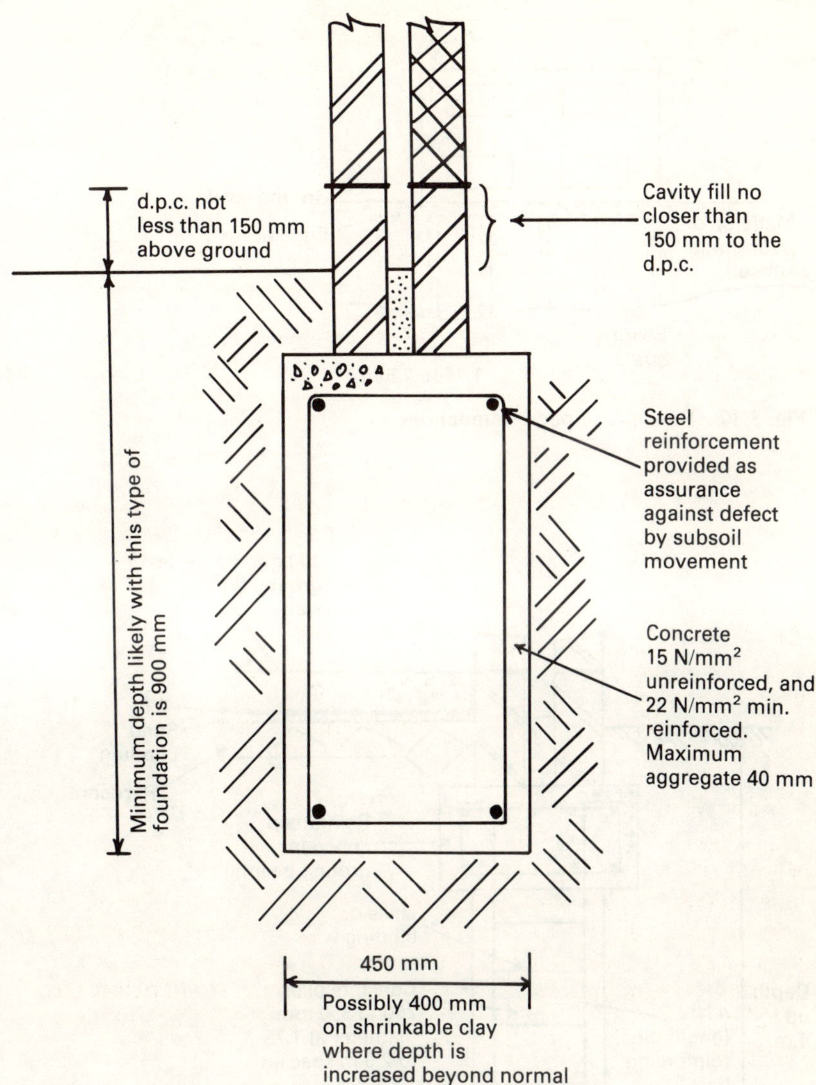

Fig. 5.16 Reinforced deep strip foundation

load-bearing walls. This is a noticeably effective method in poor soils and compares economically with conventional techniques in similar situations. See Fig. 5.18 for application of this technique.

(b) *Bored and under reamed*
Boring methods are similar to that of end-bearing piles, but to a predetermined depth depending on the nature of the subsoil.

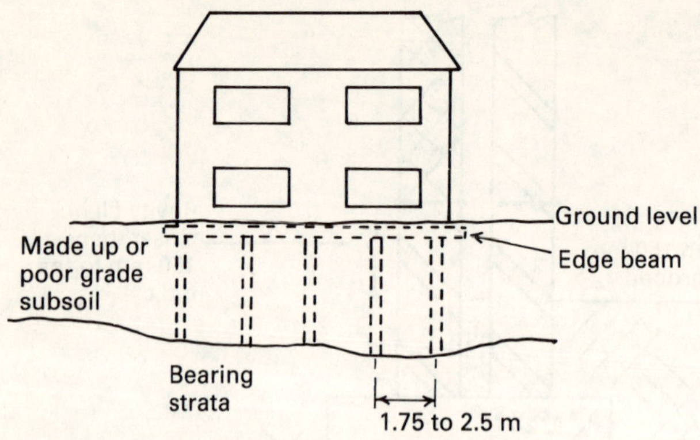

Fig. 5.17 Principle of piled foundations

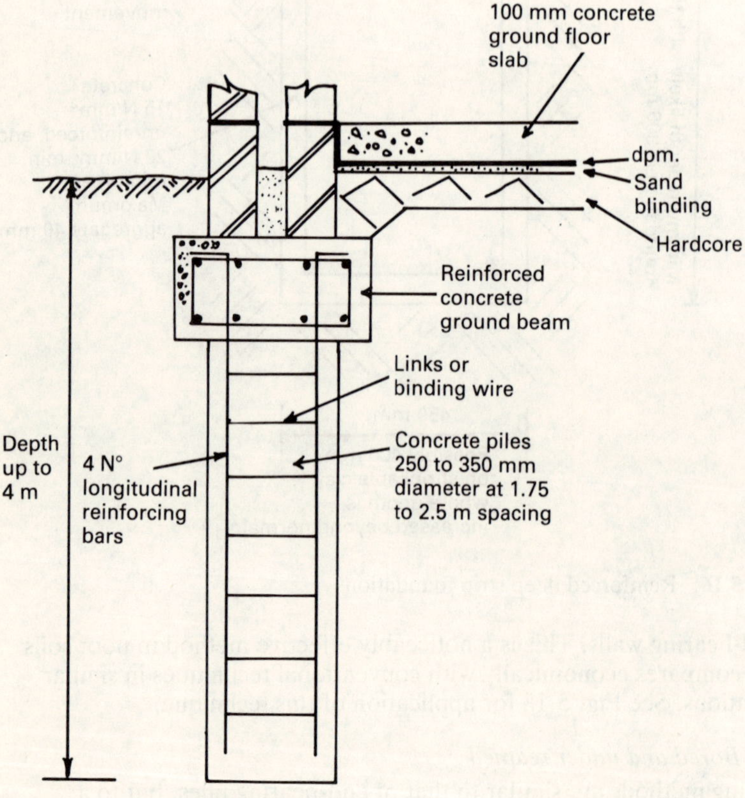

Fig. 5.18 Bored and cast *in situ* reinforced concrete pile foundation

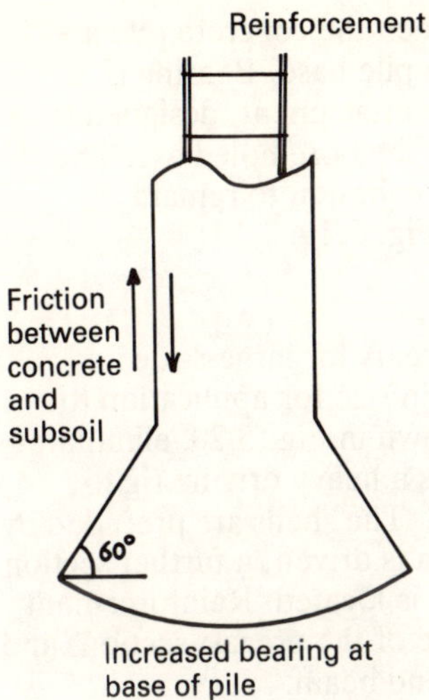

Fig. 5.19   Under reamed pile foundation

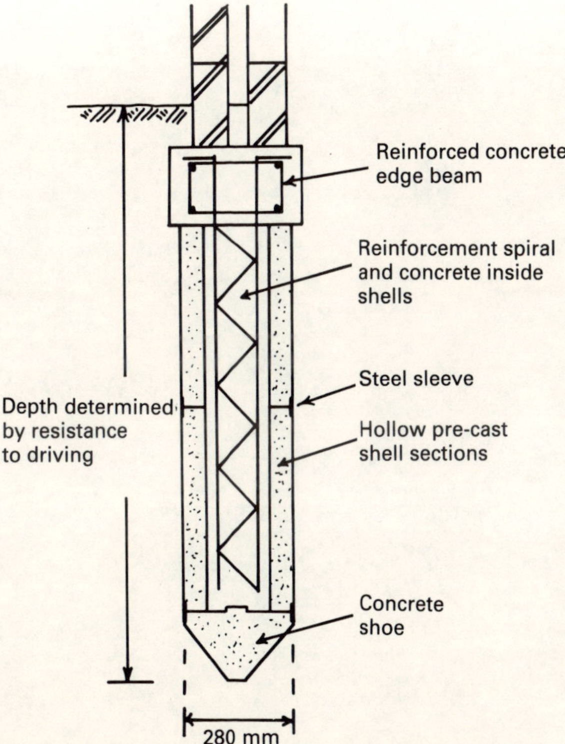

Fig. 5.20   Pre-cast driven mini shell pile foundation

Resistance to settlement is by friction between the concrete pile and surrounding subsoil and bearing area of the pile base. Bearing is increased by use of an under reaming auger attachment, designed to open and excavate into subsoil surrounding the bore hole base. The technique is limited to subsoils of sufficient cohesion to remain undisturbed during boring and is shown in Fig. 5.19.

(c) Driven pre-cast concrete piles

This method has been employed for many years for large-scale and marine construction, but has been reduced in size for application to sites of low-grade subsoil. This method, shown in Fig. 5.20, eliminates the need for boring equipment, but requires a heavy driving rig to 'hammer' the pre-cast shells into the ground. The shells are preceded by a pointed concrete shoe, and as each section is driven, a further section is placed to follow until sufficient resistance is located. Reinforcement and concrete are placed in the hollow centre of the precast sections and linked by conventional *in situ* concrete ground beam.

Chapter 6

Concrete — application to ground level floor slabs and foundations

Mixing

Concrete will be required in varying quantities for foundations, ground floor slabs, paving and driveways. The decision to use ready-mixed supplies or site mixing will be determined primarily by the programme of work. Other factors such as special mixes and quantity of mixed concrete, in addition to site accessibility, are also to be considered. See Fig. 6.1 for a guide to ready-mixed concrete lorry dimensions as an indication of accessibility.

Ready-mixed concrete

Ready-mixed concrete cannot be produced as quickly and conveniently as site mixing. In similarity with other building materials, it must be ordered in good time, and to avoid confusion the precise components must be indicated. The form shown in Table 6.1 could be used for this purpose. Co-ordination of deliveries must be agreed with the supplier, otherwise ready-mixed concrete lorries will be kept waiting, with the concrete setting process commenced! At the other extreme, site operatives will be unproductive between deliveries. Nevertheless, if the character of the work is simple and variations in deliveries are unlikely, the savings in construction time by far offset the additional purchasing cost of ready-mixed concrete. A further advantage over site mixing is the guaranteed quality of the mix, less easily achieved with site mixing.

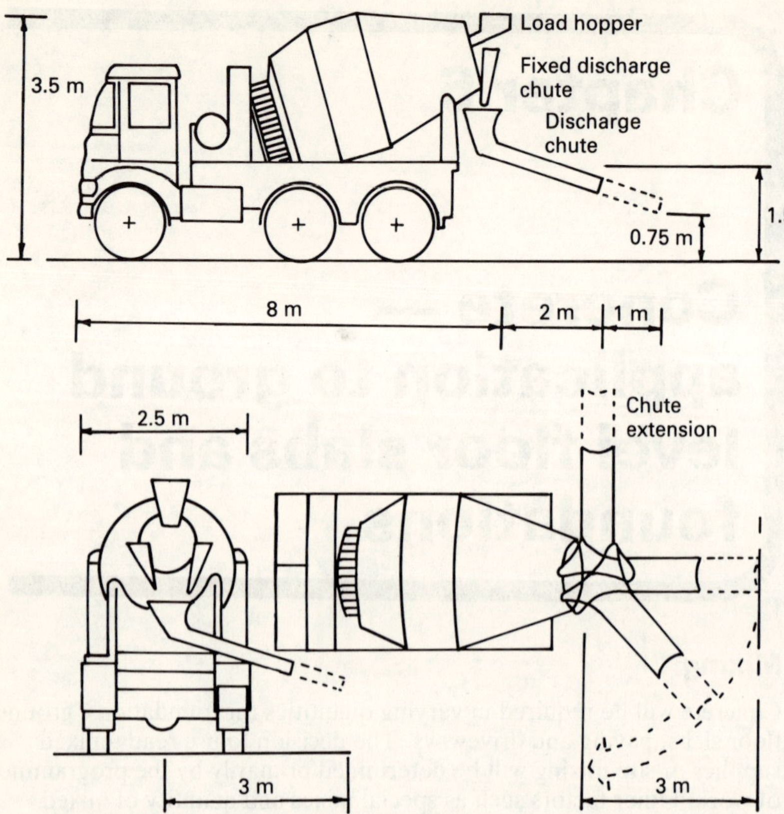

Fig. 6.1 Ready-mixed concrete lorry and dimensions

Table 6.1

Details of purchaser
Name and address of contractor ....
Address of site ....
Contact on site ....
Telephone number, office .... site ....

Details of concrete
Date required ....
Amount required ....
Grade (see Ch. 12) ....
Type of cement .... Minimum content ....
Type and size of fine aggregate ....
Type and size of coarse aggregate ....
Maximum aggregate size ....
Workability/slump ....

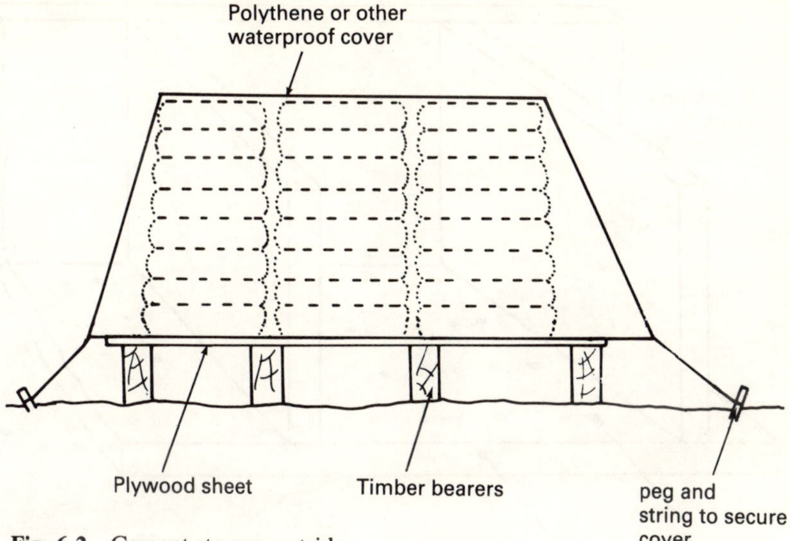

Fig. 6.2 Cement storage outside

Site mixing

Storage of aggregates and cement
For mix reliability it is essential that concrete components are correctly
stored.

Cement   The majority of housing sites do not justify the use of silos
for cement storage, therefore dry storage facilities for the standard
50 kg bags is essential to prevent dampness causing a reaction with the
cement. The minimum effective cover is a base of timber bearers and
boarding for the bags to clear ground dampness, and a tarpaulin or
plastic sheet secured over the bags, similar to that shown in Fig. 6.2.
An improvement is a shed or surplus site hut, with flooring clear of the
ground. Here, bags must not touch the sides as dampness easily
penetrates the thin plywood cladding.
   Whether stored outside or within a shed, cement should be used in
date rotation, as the longer cement is left the more it will absorb
moisture from the atmosphere.

Aggregates
   Aggregates are not damaged by moisture, but the water content
retained in the voids between sand and gravels will have a considerable
affect on the water/cement ratio of the mixed concrete, unless
allowance is made for this condition. Storage should be on a clean hard
surface, preferably concrete with small drainage perforations. A
substantial base will prevent aggregate wastage, and impurities such as

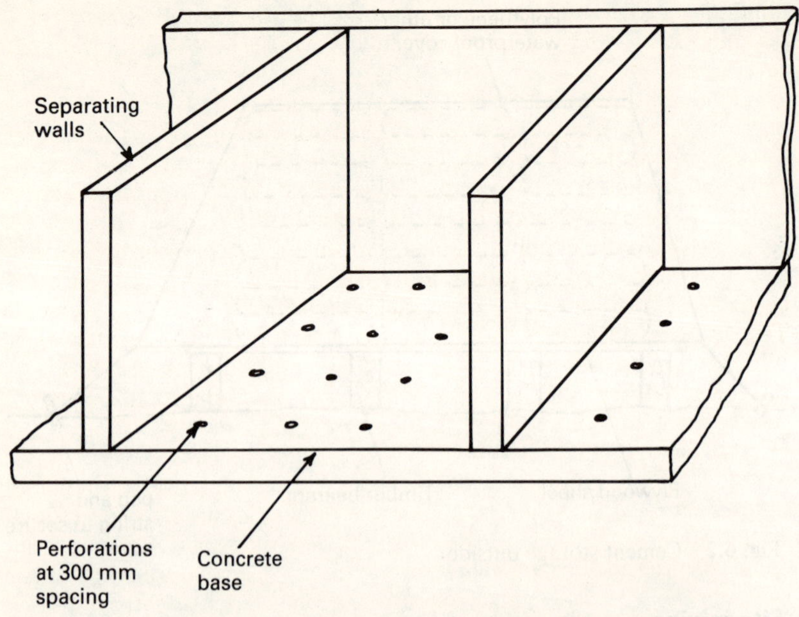

Fig. 6.3 Storage of aggregates

mud and leaves contaminating the mix. Where variable mixes are specified, separation of different graded aggregates is essential. The method shown in Fig. 6.3 is ideal, using temporary dividing walls of blocks or timber sheeting.

Site mixers

Three types of portable concrete mixer suitable for application to estate and house construction are shown in Fig. 6.4. These are the small tilting drum (½ bag) mixer, frequently used for mortar mixing, in addition to producing small batches of concrete and the two variations of the larger capacity non-tilting mixer.

Tilting mixers are powered by petrol or diesel, with small versions driven by electricity. The motor, mounted at one end, drives the electricity. The motor, mounted at one end, drives the mixing drum at approximately 20 r.p.m. The drum may be locked in various positions by the hand-operated wheel, located at the opposite end to the motor. Capacity varies, and models range from between 75 l of mixed concrete to 200 l.

Manufacturers designate these mixers by the capacity unmixed relative to mixed, e.g. 150/100 T. This means that 150 l of dry aggregate and cement in the drum will reduce to 100 l when water is introduced. Water has the effect of diluting the finer materials and filling the voids. T indicates tilting drum. Non-tilting mixers have the designation NT or NTR following their capacities, the R signifying reversing drum.

The non-tilting pattern has a rotating drum mounted on a horizontal axis. Inside and attached to the drum are blades to assist in mixing the concrete. Concrete extraction is by introducing a hinged chute into the path of the rotating concrete, redirecting it out of the drum. The reversing drum mixer contains blades arranged to screw the concrete drum mixer contains blades arranged to screw the concrete towards the front of the mixer, and then throw it back. The cycle continues until thorough mixing, when the rotating action is reversed. The screw blades then direct the concrete to the discharge orifice at the opposite end of the drum, to the inlet. The majority of non-tilting mixers have larger drums than tilting mixers, and mixed capacities range between about 200 and 500 l.

The reversing drum principles also apply to ready-mixed concrete lorries. The lorry-mounted drum is angled at 15° to the horizontal and contains similar screw blades to the reversing drum mixer. The action is the same, but reversal of the drum delivers the concrete through the same opening which initially received the unmixed materials. Placing of concrete is by discharge chute which extends up to 3 m from the rear of the machine.

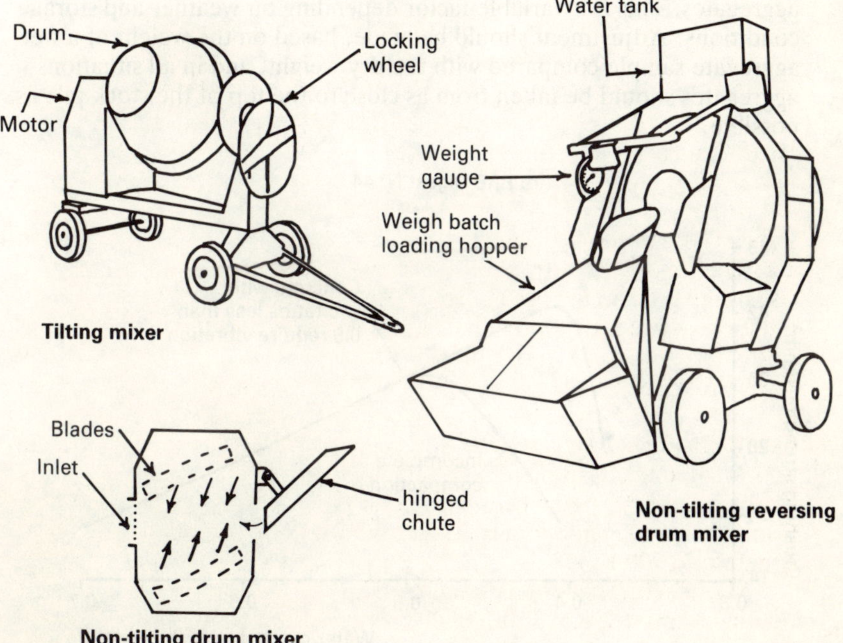

Fig. 6.4 Concrete mixers

Batching

Volume batching

Volume batching is noticeably inaccurate because of the baulking effect of fine and coarse aggregates. The inaccuracies are more obvious with large volumes, therefore, for high quality concrete, this method is only suitable for small batches. Additionally, it is a technique requiring manual gauging of quantities, a slow process, unsuitable for large-volume production.

Weight batching

Batching by weight eliminates the inaccuracies previously mentioned and considerably increases the efficiency of mixing time. The loading hopper is linked to a weighing dial for simple gauging of the contents. Water is best retained in a tank above the mixer, but use of a hose pipe is possible. Tanks contain between about 50 and 70 l, and should be provided with a contents gauge for accurate control of the very important water/cement ratio. See the graph in Fig. 6.5, where w/c ratio is seen to have a significant effect on concrete strength.

Inaccuracies are possible when daily cleaning of the mixing machinery is forgotten. Layers of cement grout or concrete build up and accumulate on the delivery chute of the hopper. Errors are also possible where allowance is not made for the water content in aggregates. This is a variable factor depending on weather and storage conditions. Adjustment should be made, based on the weight of a wet aggregate sample compared with the dry weight, and in all situations aggregates should be taken from as close to the top of the stock pile as possible.

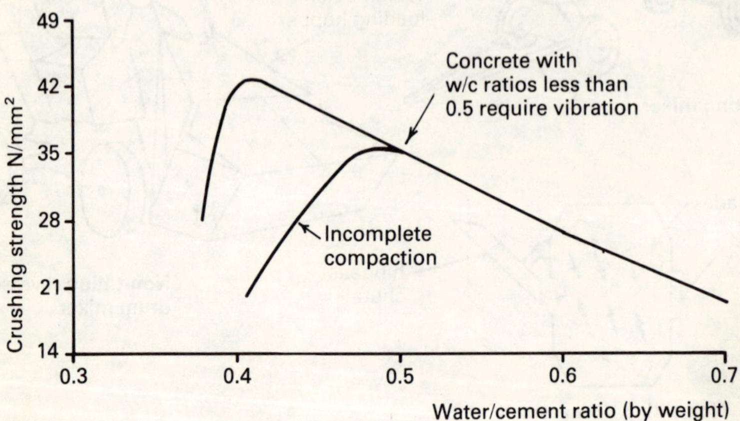

Fig. 6.5 Effect of water cement ratio on concrete strength

Transporting concrete

The mobility and chute range of ready-mixed concrete lorries is an
enormous advantage over static site mixers. Considerable time is saved
by the convenience of direct placing. Unfortunately there have been
many instances where pressurised site managers, anxious to save time,
have directed lorry drivers too close to excavations or poor ground. A
fully-laden lorry may exceed 24 tonnes, and the resulting catastrophe
of lorries 'bogged' down to the axle with trenches collapsed is an
embarrassment to be avoided!

Location of service pipes must also be considered, as they are
easily fractured when subject to loading of this nature. Transportation
by crane using hook and skip is ideal, but for housing development in
this country it is a rare sight. The irregularity of site plan will not
always lend itself to fixed plant, but mobile cranes of the type shown in
Fig. 6.6 are very useful for distribution of concrete.

Pumping is a very useful technique when site accessibility is a
problem. Efficiency is dictated by the distance the delivery pipe extends
and the resistance to flow offered by bends and dips. Lorry-mounted
pumps of the type shown in Fig. 6.7 are preferred for ease of mobility
and are designed to work in conjunction with ready-mixed concrete

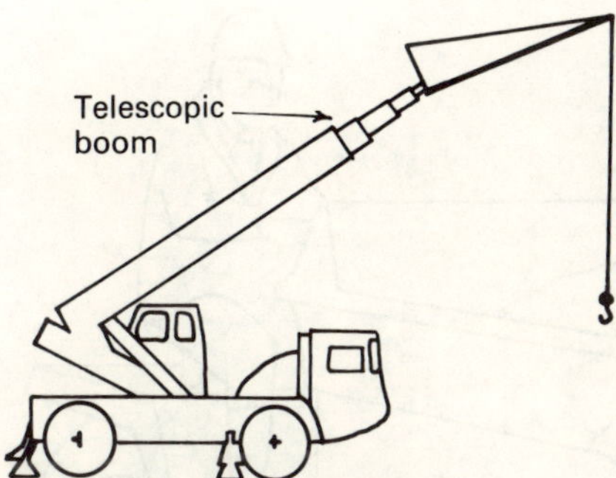

Fig. 6.6 Mobile site crane

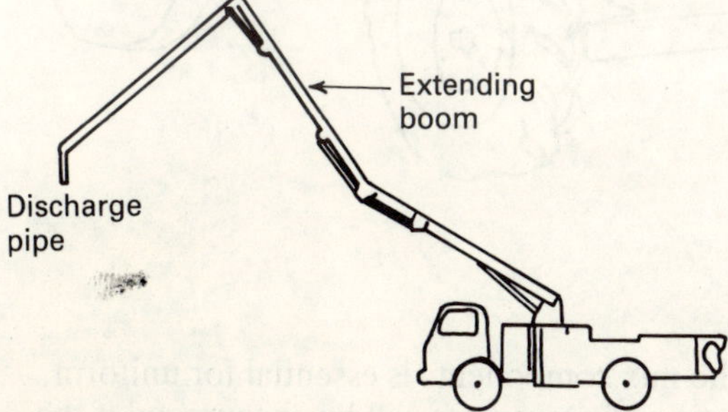

Fig. 6.7 Lorry-mounted concrete pump

lorries. The pump driven by the lorry engine contains a hopper to receive the concrete and a pair of pistons operating side by side. The pistons alternately draw the concrete in, and extrude it through the delivery tube. For effective pumping, the concrete mix specification should contain a larger than normal proportion of fine aggregates with rounded coarse aggregate. Water/cement ratio will be relatively high to ease pumping action.

Cranes and pumps are expensive to hire, and the majority of house building contractors will rely on the 'workhorse' of building sites – the site dumper, illustrated in Fig. 6.8. These are available in two- or four-wheel drive, usually with rear wheel steering and the possibility of adaptation to road use. Various sizes of machine are available, with skip capacities ranging between 0.3 m^3 and 3.0 m^3 for normal site use. Some models are ideal for concrete transport, particularly those with a rotating skip. The average builder will not have a range of machines to depend on, so concrete mixer capacity should be selected to be within that of the dumper. A pair of small dumpers are often more advantageous than one of larger capacity, for reasons of ease of access and capacity which will correspond with the work rate of the concreting gang.

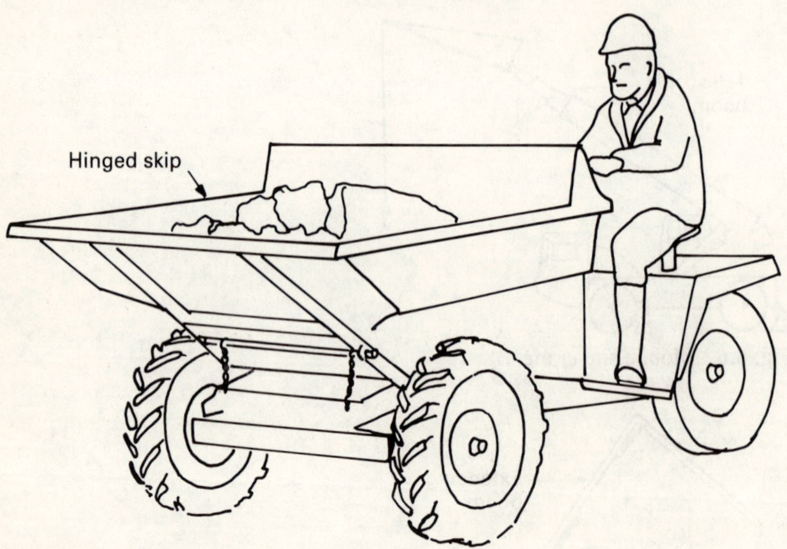

Fig. 6.8 Site dumper

Placing and compaction

The even distribution of the mix components is essential for uniform concrete strength. Segregation of aggregates will be encouraged if the wet mix is allowed to flow from the place of deposit to final placing.

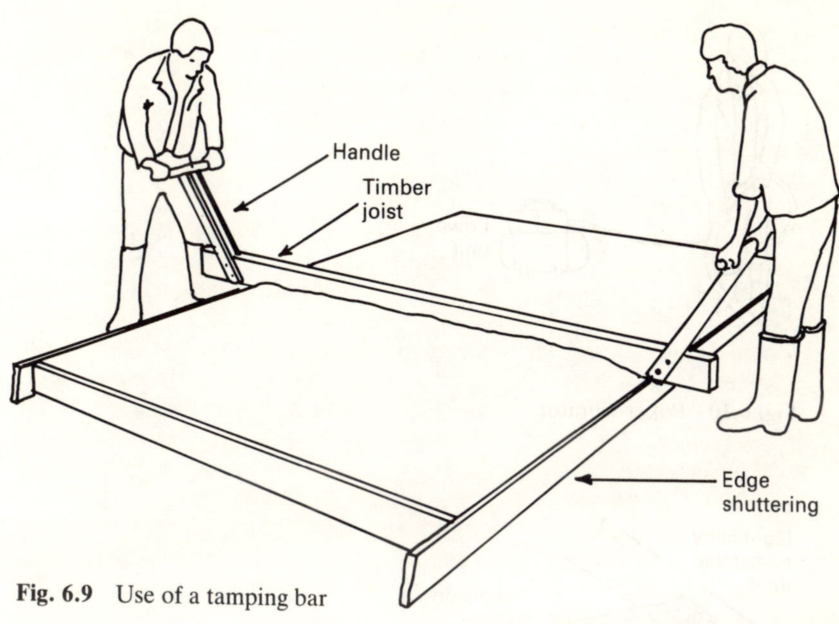

Fig. 6.9 Use of a tamping bar

Concrete should be placed and compacted as close to its final position as possible. Other defects due to aggregate segregation include variations in the finished colour, known as 'patchiness' and uneven surface texture which will not weather or wear uniformly.

Compaction is necessary to exclude trapped air and subsequent areas of weakness from the concrete. For simple ground floor slabs and garage bases the effect of a tamping bar will be sufficient. The use and construction is shown in Fig. 6.9, where finished concrete level is established from the perimeter brickwork or timber edge shuttering. For deeper concrete, such as that encountered in deep concrete strip foundations, an internal vibrating poker, driven by compressor or small petrol/diesel motor, is the only satisfactory method of air extraction. Figure 6.10 shows the use of a poker vibrator and power unit for employment in this situation. Depth and spacing of poker must be regular to ensure uniformity. Over-vibration must be avoided, as this will encourage segregation.

Finishing

The finish left by a tamping bar is often inadequate, as surface irregularities will cause uneven wear. Hand finishing with a steel trowel is impractical as accessibility to the full area of concrete is difficult if not impossible. Access by float is possible by using a light alloy adjustable float or blade attached to an extension handle as shown in Fig. 6.11. Immediately following compaction of concrete, the float is applied by pushing and drawing over the wet surface.

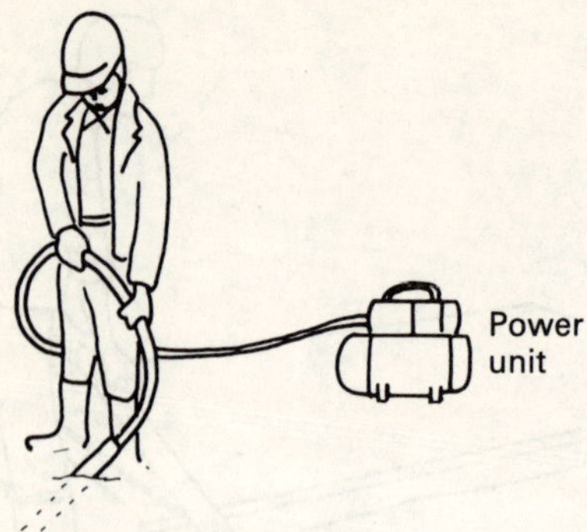

Fig. 6.10 Poker vibrator

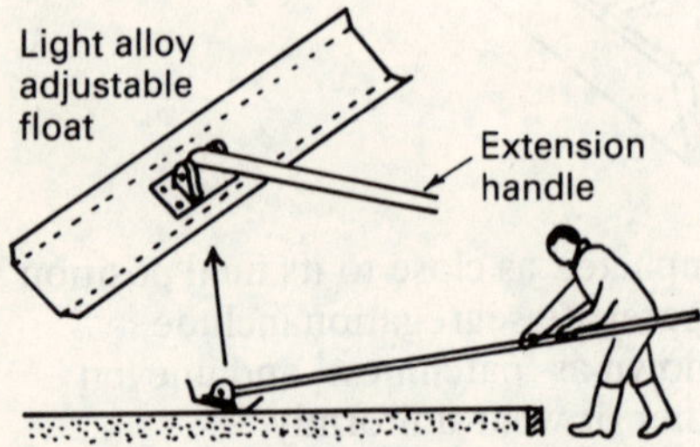

Fig. 6.11 Float-finishing concrete

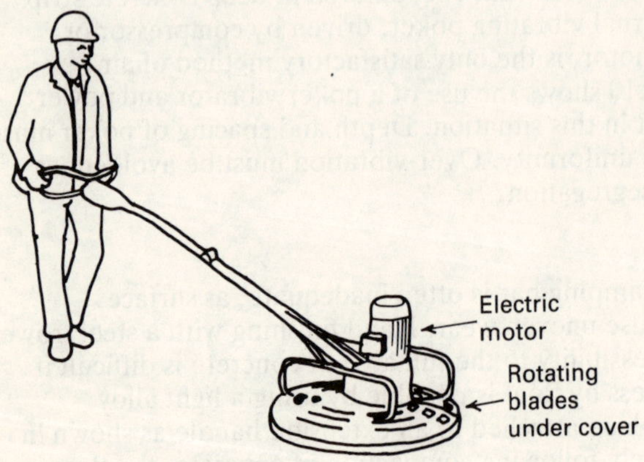

Fig. 6.12 Power float concrete finisher

The power float is an alternative, gaining in popularity over recent years. It is an ungainly looking device, as may be seen from Fig. 6.12. The motor, either electric or petrol, is mounted on a frame with a circular safety guard protecting the user from the four rotating steel blades. Model range provides overall diameters of 610, 840 and 1200 mm. An automatic cut-out throttle is fitted to an extension handle and variable control positions provide blade rotation of between about 50 and 100 r.p.m.

A high quality finish is possible, eliminating the need for screeding. This is only possible if the concrete set is sufficiently advanced and firm enough to support the operator's weight. As a rough guide, if pressing the palm of the hand onto the surface permits concrete adhesion, it is too wet for fine finishing. Rough finishing can commence if the concrete is hard enough to walk on without deep imprints occurring. Blade angles are adjustable for different effects and various attachments are available. A plate containing carborundum stones is an ideal means of polishing the surface to establish an exposed aggregate finish, or for removing surface defects and ridges. A textured finish is also possible using a scarifying attachment and this is suitable where a good key is required for screed bonding. Corners are inaccessible with this circular machinery and must be treated or finished by hand tools.

Curing

Concrete gains strength with age, attaining 75 per cent of design strength after seven days and full design strength at 28 days. Accelerated and retarded hardening are problems usually associated with the effect of weather extremes, but this is not the only cause of defective concrete. Use of 'stale' cement encourages rapid setting, manifested by cracking within the concrete and surface crazing.

Low temperatures will prolong the setting and hardening time indefinitely, and during the winter season it is advisable to use additives for accelerating the set to protect the concrete from frost action. Rapid-hardening Portland cement is a possibility in these conditions; it is very finely ground and produces a concrete of high strength in approximately half the time taken by ordinary Portland cement concrete. In very cold weather the strength gain period could extend to three quarters of the time.

Over exposure to the sun and wind will cause too rapid evaporation of the moisture content of concrete. The immediate effect is chalkiness or a powdery finish to the surface, but deeper than this small cracks may develop, appearing superficially as crazing.

Curing methods are numerous; traditional techniques include the use of wet sand or damp sawdust spread over the surface. Problems of surface disruption, staining and uneconomic use of labour eliminate their use today. Spraying with water is a sound practice, obviously only applicable when the concrete has achieved an initial set. Water should be applied in spray form at about the same temperature as the surface and care must be taken to ensure that surface drying between spraying is avoided.

Hessian sheeting soaked in water is also very successful. It should be laid when the concrete is sufficiently hard enough to avoid surface damage. This form of sheet curing is now virtually superseded by the use of waterproof paper or polythene sheeting. These have the benefit of retaining the moisture content and allowing full hydration. Light coloured sheeting will reflect the sun's rays in warm weather, but black is preferable in cold weather because of its heat absorption qualities. The only disadvantage of waterproof sheeting is its tendency to encourage uneven distribution of surface water and condensation. This could create discolouration and surface texture imperfections.

Concrete curing membranes applied by spraying are based on waxes, resins or chlorinated rubber. These provide a quick-drying, water-impervious film of up to 90 per cent efficiency, and are applied when surface water and sheen have just disappeared. If the surface is left for too long, it will absorb the spray and cause problems with the finish. Aluminium or white pigmented additives are possible for effective reduction in light and heat absorption. As a precautionary note, these compounds should never be used where surface bonding to screeds or additional concrete is required. The exception is with special grades, which disintegrate and break down after about 30 days of application. Removal is then by stiff brushing.

Solid concrete ground floor construction

The Building Regulation requirements for solid concrete ground floors bearing directly on the subsoil are found in section C3. This contains no precise specification of materials or dimensions, unless continuing to section C4, where the requirements for oversite concrete to suspended timber floors may be deemed satisfactory. For these purposes concrete is a minimum of 100 mm thick, of 1 : 3 : 6 mix proportions and laid on a clean bed of hardcore. Regulation C3 insists that the floor contains means of preventing moisture rising from the ground to the upper surface; this is normally by incorporating a damp proof membrane into the construction. Furthermore, the material composition of the floor must be of a quality which will not be impaired by ground moisture. Any hardcore or filling used under the floor must be free from sulphates and other materials likely to encourage deterioration of the floor structure.

For more practical consideration of legislation governing solid ground floor construction, reference may be made to the National House Builders Council handbook of technical requirements. The illustration in Fig. 6.13 incorporates many of these requirements in addition to the Building Regulations. The NHBC require the concrete

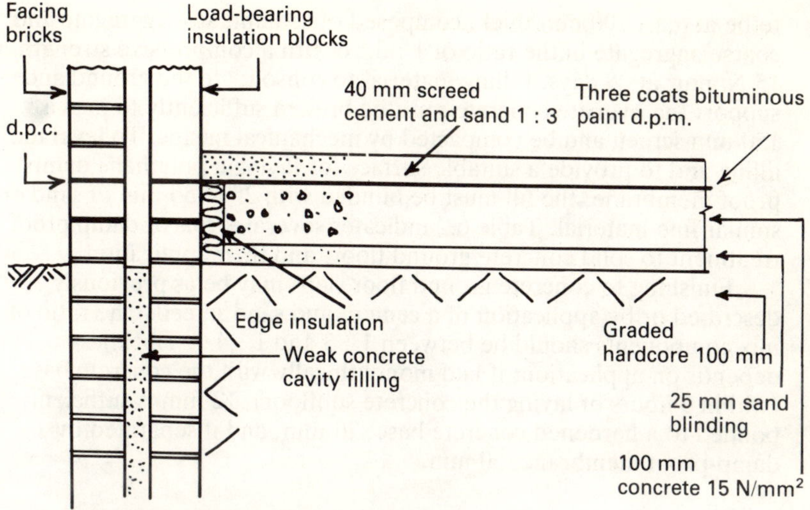

Fig. 6.13 Solid concrete ground floor construction

Table 6.2 Table to NHBC rule Fo 26(a)
Damp proof treatment to solid concrete ground floors

In position below the slab	In sandwich construction	On the surface of the slab
Polythene film, not less than 500 gauge (0.13 mm)	Any sheet material given in previous column	Hot applied mastic asphalt or pitch mastic
Bitumen sheet to BS 743, *Materials for damp proof courses*	Hot applied mastic asphalt, pitch, or bitumen	Cold applied pitch epoxy resin
	Cold applied bituminous solutions, coal tar pitch or rubber emulsion applied not less than 0.6 mm thick	
	Composite polythene and bitumen self adhesive of not less than 0.6 mm thick	

to be at least 100 mm thick, composed of cement, fine aggregate and coarse aggregate in the ratio of 1 : 3 : 6 with a compressive strength of 15 N/mm^2 at 28 days. Filling material to consolidate the ground and support the oversite concrete must be broken sufficiently to pass a 150 mm screen and be compacted by mechanical means. To level the filling and to provide a suitable surface for laying a polythene damp proof membrane, the fill must be blinded with 25 to 50 mm of sand or similar fine material. Table 6.2 indicates several forms of damp proof treatment to solid concrete ground floor, and their application.

Finishing to concrete ground floor slabs may be as previously described or by application of a cement and sand screed. The ratio of mix components should be between 1 : 3 and 1 : 4.5. Thickness depends on application; if laid monolithically with the concrete base (within 3 hours of laying the concrete subfloor), 12 mm is sufficient; if bonded to a hardened concrete base, 20 mm, and if separated by a damp-proof membrane, 50 mm.

Chapter 7

The function of external and internal walls

External walls

Functional requirements
1. Strength and stability
2. Weather exclusion and durability
3. Thermal insulation
4. Sound insulation
5. Fire resistance
6. Appearance

Strength and stability

The materials employed for the external wall construction of
conventional residential buildings are bricks, stones or concrete blocks.
The specification of brick and block sizes are considered later, but for
purposes of wall strength the most acceptable brick materials are burnt
clay, sand and lime (calcium silicate) and concrete. Square dressed
natural stone is also widely accepted. Where bricks or blocks are
manufactured with voids or indentations, the aggregate volume of solid
material must be no less than 50 per cent of the overall volume. Voids
created by aeration of concrete are deemed nonexistant, the material
being treated as solid. The Building Regulations require the
compressive strength of bricks and blocks used for external wall
construction of a dwelling to be a minimum of 5 N/mm^2 and
2.8 N/mm^2 respectively.

    The strength of mortar jointing is often overlooked. If this is to

Table 7.1 Building Regulations, table to rule 11.
Minimum thickness of certain external walls, compartment walls and
separating walls.

Height of wall	Length of wall	Minimum thickness of wall
Not exceeding 3.5 m	Not exceeding 12 m	190 mm for the whole of its height
Exceeding 3.5 m but not exceeding 9 m	Not exceeding 9 m	190 mm for the whole of its height
	Exceeding 9 m, but not exceeding 12 m	290 mm from the base for the height of one storey, and 190 mm for the rest of its height
Exceeding 9 m but not exceeding 12 m	Not exceeding 9 m	290 mm from the base for the height of one storey, and 190 mm for the rest of its height
	Exceeding 9 m but not exceeding 12 m	290 mm from the base for the height of two storeys, and 190 mm for the rest of its height

function effectively it should be composed of 1 part Portland cement, 1
part calcium lime and a maximum of 6 parts fine aggregate. Other
mortars are acceptable, providing they are of equal strength or greater
if the situation justifies it. All proportions are measured by volume
when materials are dry.

Stability is a factor determined by three variables:

1. The height of the wall
2. The length of the wall
3. The thickness of the wall

Table 17.1 shows the table to Rule 11 extracted from the Building
Regulations for the inter-relationship and significance of these three
factors. Notice that the width dimensions are for solid brick or block
wall construction; for stone walls the minimum thickness is to be $1\frac{1}{3}$
times that shown. For cavity wall construction the sum of the thickness
of the two leaves and 10 mm shall be not less than the dimension
specified in column 3.

The leaves of each wall used in cavity construction are not less
than 90 mm in thickness, and tied together with ties of a standard
acceptable to BS 1243 : 1978. Figure 7.1 illustrates ties suitable in this
situation and Fig. 7.2 shows the mandatory maximum spacing of
900 mm horizontally, 450 mm vertically and staggered. 300 mm

vertical spacing is required at openings for doors and windows. The cavity width is normally a minimum of 50 mm and a maximum of 75 mm. In circumstances where vertical twist type wall ties are used and spaced at 750 mm horizontally, the cavity may be increased to 100 mm.

Wall thickness must not be less than $\frac{1}{16}$ the storey height in which it is contained and no less than any other wall it supports. The exception to the fraction ruling is where small buildings and annexes have external walls not subject to cavity construction. For example, a garage or porch has a minimum acceptable wall thickness of 90 mm provided the wall height does not exceed 30.0 m. If either the height or length exceed 2.5 m, stability is provided at intervals of 3.0 m maximum with attached piers of at least 190 mm square sectional area. Figure 7.3 represents a simple interpretation of these requirements.

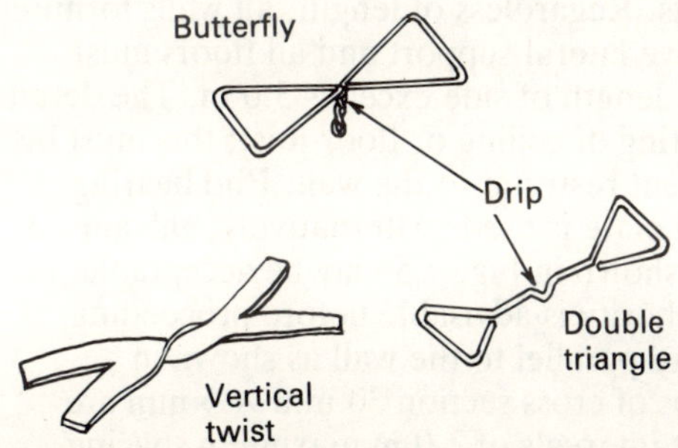

Fig. 7.1   Galvanised wall ties to BS 1243

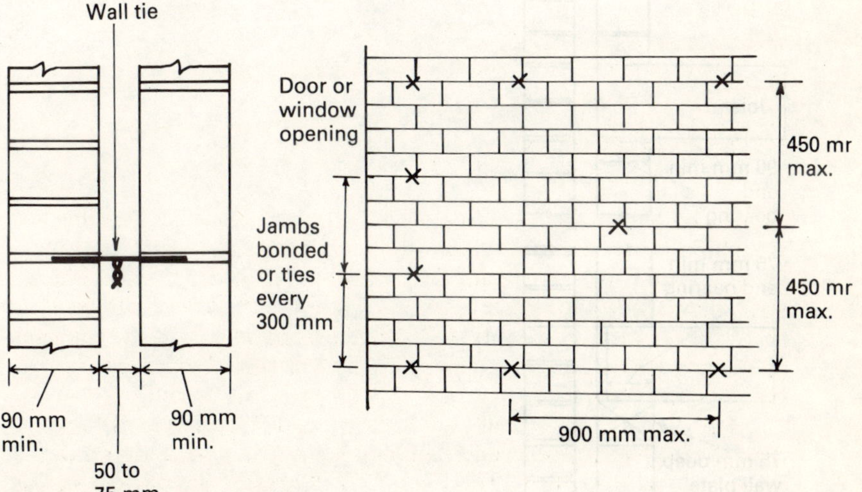

Fig. 7.2   Spacing of wall ties

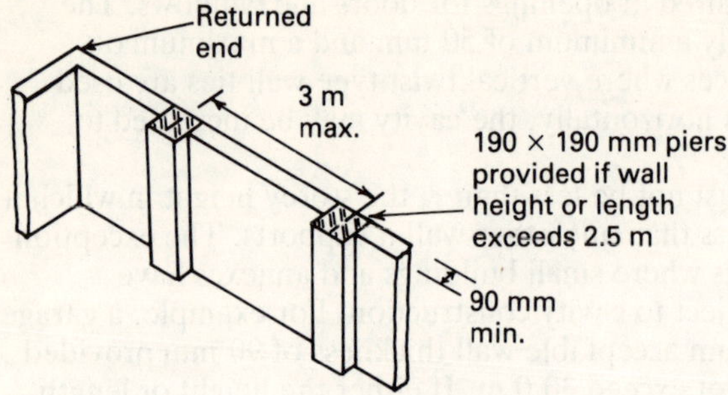

Fig. 7.3   Stability of half brick walls

Additional requirements for the stuctural stability of cavity walls includes the use of steel strapping to provide lateral support from adjacent ceiling and floor joists. Regardless of length, all walls forming a junction with a roof must have lateral support and all floors must provide lateral support if their length of side exceeds 3.0 m. The detail in Fig. 7.4 represents end bearing of ceiling or floor joist; this must be at least 90 mm to apply sufficient restraint to the wall. End bearing may reduce to 75 mm if a wall plate is used. Alternatively, galvanised steel joist hangers of the type shown in Fig. 7.5 may be acceptable, but consultation with the local authority is advisable before proceeding with these. Where the joists are parallel to the wall as shown in Fig. 7.6, galvanised steel straps of cross section 30 mm × 5 mm are required to anchor the floor at intervals of 2.0 m maximum spacing.

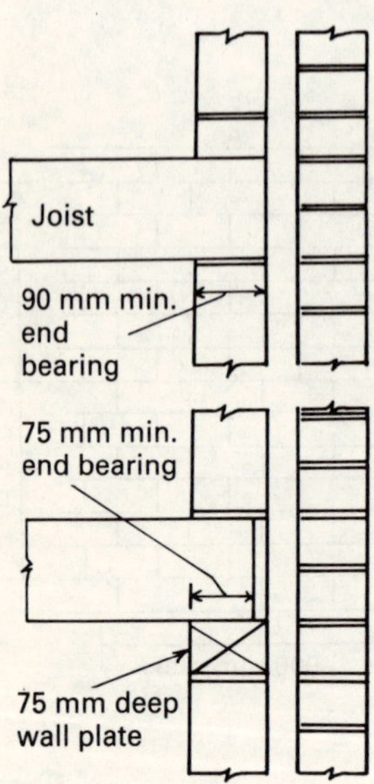

Fig. 7.4   Lateral restraint to walls

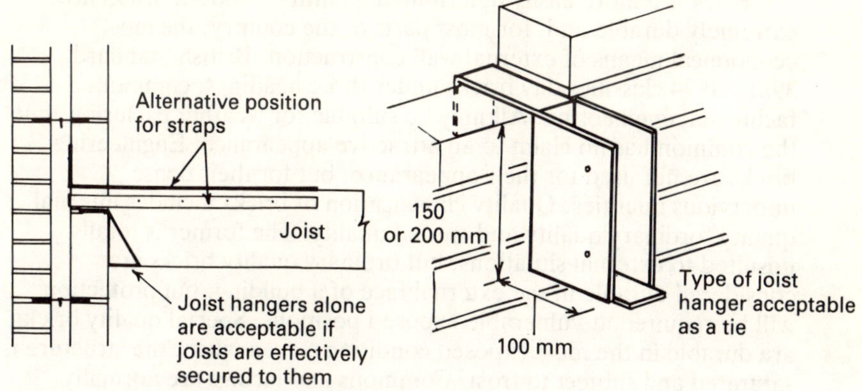

Fig. 7.5   Alternative lateral restraint to walls

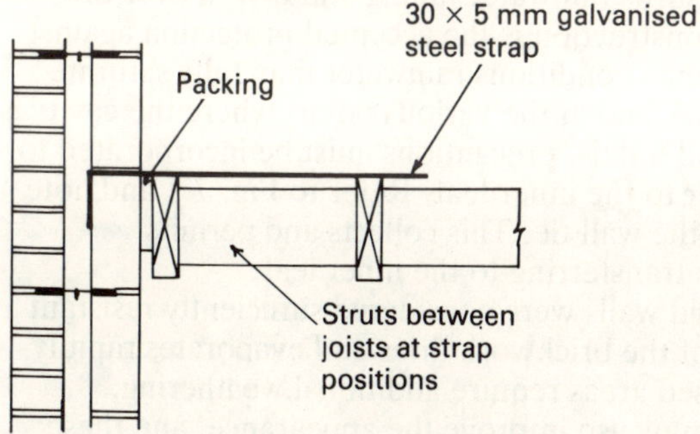

Fig. 7.6   Lateral restraint with joists parallel to wall

Weather exclusion and durability

External walls must withstand the effects of water, frost, soluble salts
and temperature change. The quality of some natural, unprocessed
wall materials such as sandstone is questionable and their weather and
chemical resistance have failed with the effects of time. Other stones,
granite and flints of dense impervious properties generally have
excellent durability. Sandstone is soft when quarried, hardening or
seasoning with exposure. The effect of frost action and absorbtion of
acids from the atmosphere often disintegrate and decompose the
porous surface. Treatment is possible, normally in the form of applied
waterproofing liquids, but care must be taken when selecting a suitable
product, as many seal the wall completely from external effects which
eliminates one problem but creates another. The wall can no longer
'breathe' and condensation may form in the porous voids leading to
internal disintegration of the stone or, with older solid walls, internal
damp patches.

Brick is a more easily determined quantity. Modern bricks are extremely durable and, for most parts of the country, the most economical means of external wall construction. British Standard 3921 : 1974 classifies day bricks under three headings: common, facing and engineering. All may be suitable for weather exclusion, but the common has no claim to an attractive appearance. Engineering bricks are not used for their appearance, but for their dense impervious qualities. Quality classification of bricks includes internal quality, ordinary quality and special quality. The former is totally unsuited to external situations, but ordinary quality bricks are considered durable in the external face of a building, but protection will be required at vulnerable exposed positions. Special quality bricks are durable in the most exposed conditions, even where the structure is saturated and subject to frost. Commons and facings are normally available in all three grades; engineering bricks are regarded as special quality only.

Facing brickwork absorbs rainwater falling and driven over the surface and cavity wall construction is the accepted protection against rain penetration. In extreme conditions rainwater may fully saturate the outer leaf of brickwork and at the various points where the cavity is bridged, e.g. wall ties and lintels, precautions must be incorporated to prevent moisture transfer to the inner leaf. Refer to Fig. 7.1 and note the drip in the centre of the wall tie. This collects and permits rainwater to drop before transferring to the inner leaf.

Traditional brick solid walls were considered sufficiently resistant to rain. Any saturation of the brickwork dries and evaporates rapidly and only the more exposed areas require additional weathering. Weathering treatments may also improve the appearance, and these include:

1. Water-repellent liquid treatment
2. Rendering with cement and sand
3. Cladding with tiles or timber

1. Water-repellent liquid treatment

Colourless water repellents are usually preferred as they preserve the natural appearance of a wall. They may be classified as:

(a) Silicone-based solutions to BS 3826 : 1969. These are invisible and claim to resist all but driving rain. The masonry pores are not sealed, which permits condensation and moisture evaporation. They are available in three categories, A, B and C, to suit most surfaces.
(b) Aluminium and poly-oxo stearates and titanates. Performance often matches the requirements of BS3826, but chemical composition differs.
(c) Resinous solutions such as polyurethane and acrylic varnishes form a colourless film having slight gloss and a darkening effect on the

masonry. Pores may block with these solutions, and flaking after a few years is quite common.

Decorative brush treatments are numerous, ranging from traditional lime- and cement-bound washes to synthetic polymer emulsions and other plastic or rubber-bound solutions. The range of colours and textures is enormous, but care must be exercised in selection and specification of these treatments as most fulfil the fundamental requirements of adhesion and waterproofing, but may fail to accommodate vapour escape from the wall.

2. Rendering

Rendering should provide a jointless, durable water-repellent finish to a wall. To fulfil this it must be free of cracks and have good background adhesion.

Choice of mix. The mortar mix should have high suction and be slightly porous and rather weak. A mix of these properties will generally contract, avoiding local cracking often associated with strong mortars. A slightly porous render will absorb some water, but insufficient to allow penetration of the background. In all but extreme conditions the amount absorbed will soon evaporate.

One application of mortar about 10 mm thick is sufficient on sheltered surfaces but elsewhere the brick or block joints will 'grin through' in damp weather. For normal applications two coats are required, an undercoat of 8 to 16 mm and a finish coat of 8 to 10 mm. In conditions of severe exposure, three-coat work is recommended. British Standard 5262 : 1976 recommends mortar mixes to suit various backgrounds and degrees of exposure. If precise details are required refer to table 1 and 2 in the British Standard. For general use a mix of 1 part cement, 1 part lime to 5 or 6 parts sharp sand will suit a sound background of bricks or blocks. Fine building sand may be substituted where a smooth finish is required.

Background. Joints of brickwork should be raked out about 12 mm to provide a key for the undercoat. Blockwork of porous materials has sufficient adhesion without attention to joints, but sufaces should be dampened to avoid loss of moisture from the rendering. Dense, smooth surfaces must be hacked to provide a key or applied with spatterdash treatment. This consists of a strong cement and sharp sand mortar in the ratio of about 1 : 2 with very little water, trowelled thinly over the surface to provide a key for the undercoat.

The surface of old brickwork may contain paint or be impractical to rake out existing joints. In these circumstances metal lathing is secured to the brickwork as shown in Fig. 7.7, but three applications of render will be required to prevent rain penetration and rusting of the lathing.

Finishes.   Numerous possibilities exist including smooth, stippled, textured and pebbledash. Smooth finishes can look patchy, but are ideal for external emulsioning. Stippled effects are produced with a coarse stiff brush and textured by application of a variety of mechanical implements. Pebbledash is a rough finish created by throwing small pebbles on to the finish coat. Coarse finishes have the disadvantage of attracting grime and dirt, particularly in town and city areas.

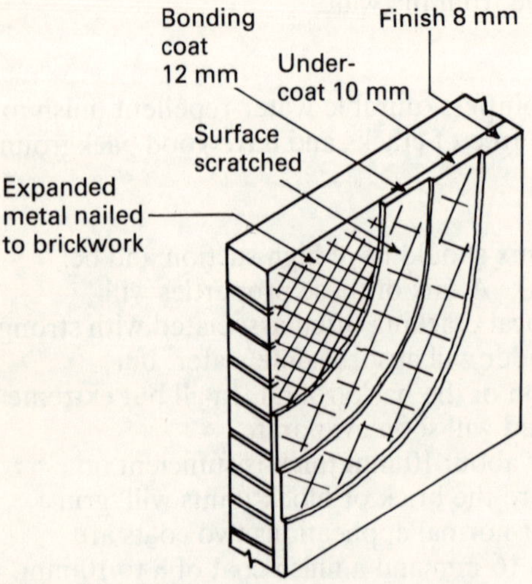

Fig. 7.7   Rendering to a difficult surface

3. *Tile cladding*

Tile cladding has been a popular facing and weathering material for several hundred years, and well suited to waterproofing existing or new walls. With new construction, lightweight concrete blocks may be situated in the outer leaf, saving in construction time and providing a considerable thermal improvement over brickwork. 32 × 19 mm tile battens may be secured direct to the wall or onto counter battens at 450 mm spacing. Counter battening, shown in Fig. 7.8, is preferred as this reduces the amount of direct nailing to the wall and it increases the air space under the tiling, therefore improving thermal insulation. For plain tile cladding the lap should be at least 32 mm. This is the amount that each tile effectively overlaps the tile-but-one below. The lap corresponds to a gauge or batten spacing of 116 mm. Plain tiles are available in various textures and colours with further variations possible with the ornamental arrowhead or club patterns shown in Fig. 7.9.

Efflorescence and sulphate attack
Efflorescence is the harmless migration of salts from within bricks to

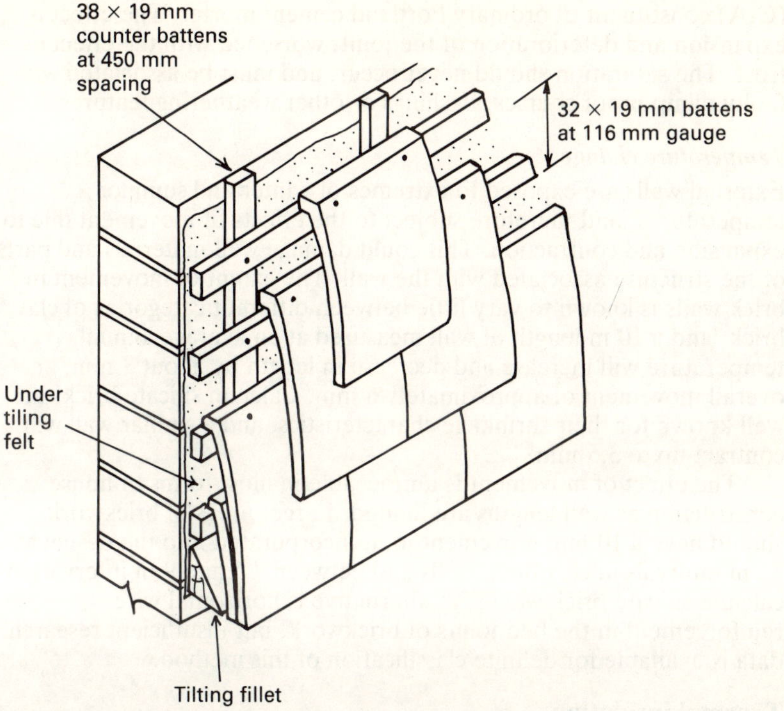

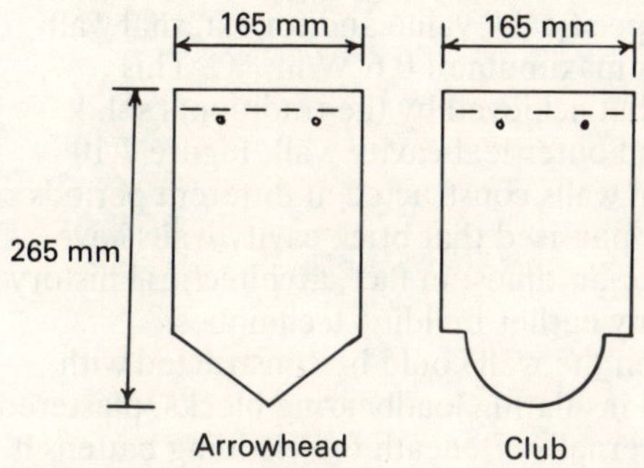

Fig. 7.8   Tile hanging to brick or block external walls

Fig. 7.9   Feature tiles

their surface. The salts originate from the clay used in brick
manufacture, and from salts contained in the ground surrounding the
wall. The effect is quite common on certain bricks, notably stocks, and
subsequent brushing will eliminate the problem.

Sulphate attack is more serious and caused where sulphates in
saturated brickwork dissolve and react with the tri-calcium aluminate

94

(C₃A) constituent of ordinary Portland cement mortar. The effect is
expansion and deterioration of the joint, worsened with the effect of
frost. The saturation should never occur, and must be associated with
faulty damp proof courses, flashings or other weathering features.

Temperature change

External walls are exposed to extremes of winter and summer
temperatures, and therefore subject to the effects of movement due to
expansion and contraction. This could damage wall materials and parts
of the structure associated with the wall. The extent of movement in
brick walls is known to vary little between different categories of clay
brick, and a 10 m length of wall measured at an average annual
temperature will increase and decrease in length by about 3 mm, an
overall movement of approximately 6 mm. Calcium silicate bricks are
well known for their shrinkage characteristics, and a similar wall will
contract up to 3.5 mm.

The effect of movement is untraceable in most forms of house
construction as wall lengths are limited. Larger areas of brickwork
should have a 10 mm movement joint incorporated into the design at
12 m intervals in clay brick walls and between 7.5 and 9 m intervals in
calcium silicate brick walls. An alternative is horizontal wire
reinforcement in the bed joints of brickwork, but insufficient research
data is available for definite classification of this method.

Thermal insulation

The mandatory requirements for minimising heat loss from the
structure are contained in Part F of the Building Regulations. Each
element of the building has a specific 'U' value and for external wall
construction this is defined as a maximum of 0.6 W/m² °C. This
standard is much higher than that achieved by the traditional solid
brick wall or the brick inner and outer leaf cavity wall. Figure 7.10
provides a comparison between walls constructed at different periods
during this century, but it is emphasised that brick cavity walls have
been in popular use since Victorian times; in fact, architectural history
reveals their application to many earlier building techniques.

To satisfy current legislation the wall could be constructed with
inner and outer leaf of thermal insulating loadbearing blocks, plastered
internally and weather clad externally. Beneath the cladding battens it
is necessary to provide a layer of waterproof undertiling felt to act as a
dampness barrier if the cladding deteriorates.

An alternative which preserves traditional appearance is
incorporation of cavity-fill insulation. The accepted materials and
techniques include:

1. Insulation slabs built into the cavity as work proceeds.
2. Rockwool loose pelleted water-repellent mineral wool filling, blown
   into the completed cavity wall.
3. Urea formaldehyde (UF) foam, cavity filling insulation.

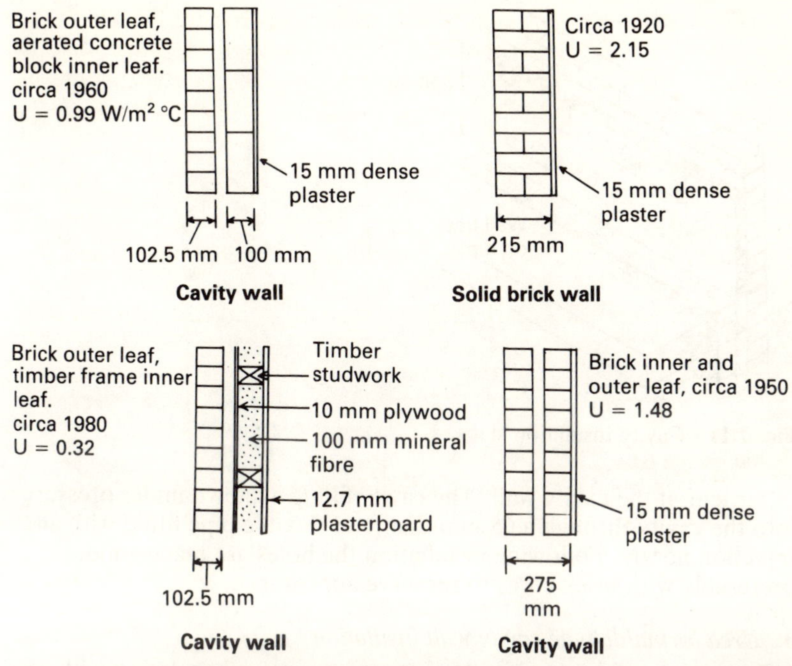

Fig. 7.10   Comparison of wall thermal insulation

Any of these provide a simple means, at modest cost, of improving the 'U' value of a conventional wall to a figure of around 0.5 W/m² °C. These methods bridge the cavity and contravene Building Regulation C9, but because of the waterproof nature of the materials, local authorities will consider relaxation of the Building Regulation for approved methods.

1. Insulation slabs

These are either rigid polystyrene slabs or water-repellent bonded glass fibre batts. They should commence as close to the damp proof course level as possible and initial support is from wall ties spaced at approximately 600 mm intervals. The slabs are manufactured slightly over 450 mm deep and are compressed between wall ties which continue at the normal spacing of 450 mm vertically and 900 mm horizontally. Close butt jointing is essential to avoid heat loss and the application is shown in Fig. 7.11.

2. Rockwool loose pelleted water-repellent mineral wool

This material may be applied to existing buildings or to new construction work, by direct filling of the cavity as work proceeds. Installation to existing buildings is conducted by boring 65 mm-diameter holes at a maximum of 2.5 m spacing, through the

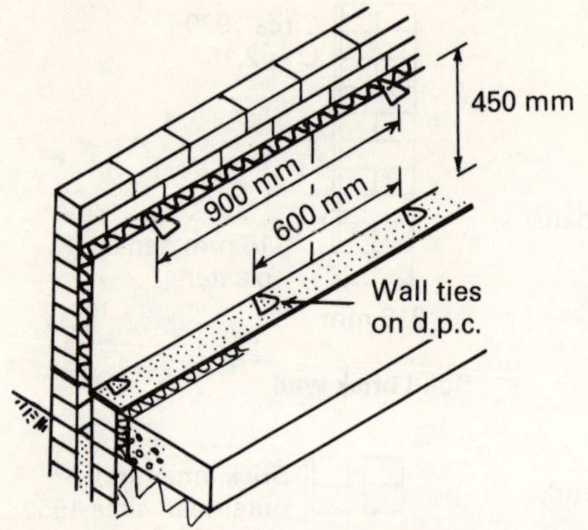

Fig. 7.11 Cavity insulation slabs

outer leaf of the cavity wall. The cavity filling is blown under pressure into the cavity through a 65 mm-diameter flexible pipe fitted with an injection nozzle. Following installation the holes are made good, preferably with brick cores to preserve appearance.

3. Urea formaldehyde cavity wall insulation

UF foam is produced by combining an aqueous resin solution with an aqueous hardener solution, air and a surfactant. The solution is pressure supplied to an injector gun and shortly after placing in the cavity it dries by fabric absorption and evaporation.

Top filling of the cavity is possible with new construction, but centre filling is the only practical method with existing structures. Centre filling requires 20 mm diameter holes bored at joint intersections in the outer leaf or inner leaf on new unplastered walls, at 1100 mm maximum spacing horizontally and 900 mm maximum spacing vertically, commencing two brick courses above the damp proof course. On completion of drilling, thin timber pegs are placed in the holes and allowed to project slightly. Injection commences at one of the lowest holes and movement of adjacent pegs signifies penetration of foam. Injection then continues from the adjacent holes and the process is repeated until full coverage is achieved. As a precaution, it is essential to ensure that the cavity is fully sealed at air bricks and other openings, particularly at the top of the cavity wall. Optimum foam density is 8 kg/m^3; below this, voids created by shrinkage could be a serious problem, encouraging water to bridge the cavity.

Sound insulation

The Building Regulations contain no specific requirements for resistance to transmission of sound through external walls. Nevertheless it is desirable and necessary to prevent traffic and other external noise sources penetrating the structure. For normal purposes the conventional masonry cavity wall is sufficient provided air paths and gaps around window and door frames are sealed.

Sound insulation between dwellings is of particular importance and this is considered in the section headed Party walls.

Fire resistance

The minimum periods of fire resistance for elements of structure are detailed in the table to Building Regulation E5. For conventional house construction not exceeding three storeys the fire resistance requirement for external walls is half an hour. Schedule 8 of the Building Regulations indicates the type of construction which satisfies this standard and conventional brick and block cavity walls are more than adequate. Reference to the schedule provides a comprehensive range of materials and their respective fire resistance. A few examples include:

100 mm clay or sand lime load-bearing brick wall = 2 hours
200 mm clay or sand lime load-bearing brick wall = 4 hours
Timber frame with $12\frac{1}{2}$ mm plasterboard on each side = $\frac{1}{2}$ hour
Timber frame with 19 mm plasterboard on each side = 1 hour

A materials ability to encourage spread of flame over its surface is represented in Building Regulation E15. Here reference to British Standard 476 : 1981 provides a classification in descending order of 0, 1, 2, 3 and 4. Restriction of surface spread of flame requirements are well satisfied by conventional brick or other masonry external-facing materials. These are of the highest class, 0, and are regarded as incombustible.

Appearance

External appearance of a wall is achieved by applied treatments such as paint, rendering and cladding or improving the exposed surface by using sanded or textured bricks. Other forms of facing brick, notably stocks, are created with the use of additives in the basic clay. These fuse with the clay during the kilning process, providing an irregular blend of colour to enhance the surface.

Joint treatment or colouring will provide many alternative features to the same brick. Mortar may be coloured by using pigments conforming to BS 1014 : 1975. Experimental panels should be constructed before the building programme commences to determine the final effect, as variables such as proportion of cement in the mix and nature of the sand will considerably affect the colour. Excessive use of pigments is to be avoided; a maximum of 10 per cent by weight of cement and maximum of 3 per cent with carbon black.

The appearance of finished brickwork may considerably change by treatment of the joint profile, but before considering the effect it is important to distinguish between jointing and pointing of brickwork.

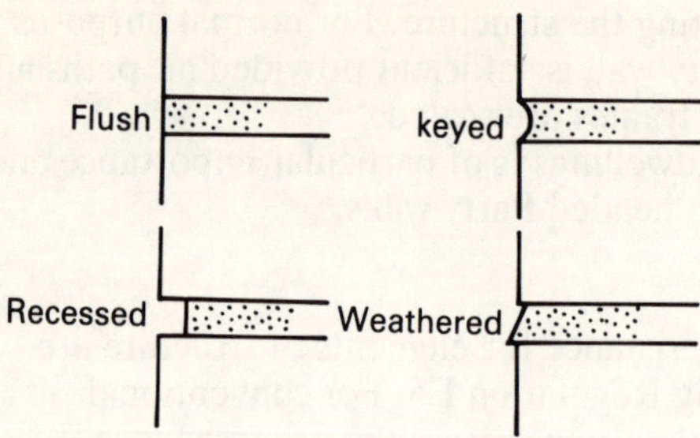

Fig. 7.12 Finishes to mortar joints in brickwork

Jointing refers to the finishing treatment of unset mortar between bricks, and pointing is raking out between 15 mm and 20 mm of unset mortar and replacing it with fresh mortar, of a particular colour or texture to provide continuity of colour. Pointing disturbs the mortar bed and reduces bricklaying output, therefore it should only be specified where a particular finish is impractical by jointing. Some possible finishes are shown in Fig. 7.12.

The size and shape of facing material will considerably affect appearance. Stone masonry walls are constructed from stones of various sizes, depending on local techniques and type of stone. Figure 7.13 provides some examples, but all possibilities are too numerous to illustrate. Brick size and shape also varies; the standard brick shown in Fig. 7.14 is 215 mm long $\times$ 102.5 mm wide $\times$ 65 mm high. The format size is often quoted and this includes a 10 mm addition for joints.

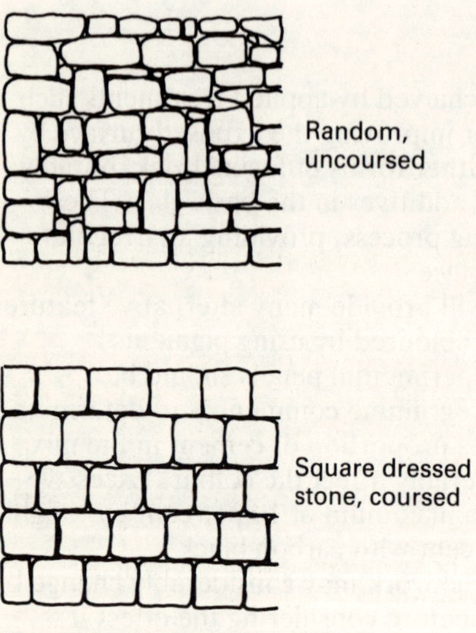

Fig. 7.13 Stone masonry

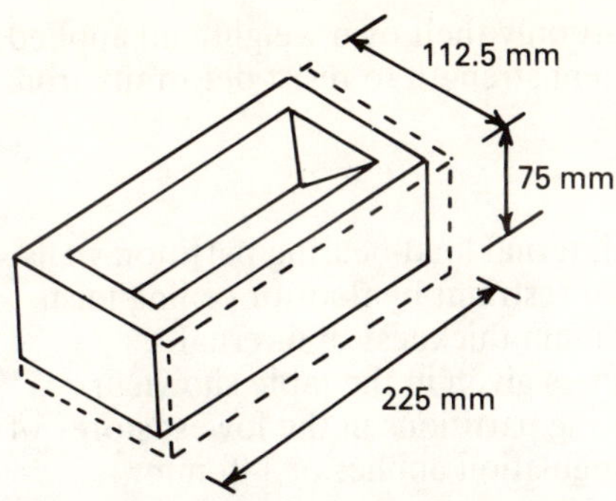

Fig. 7.14   Standard brick format size

Metric bricks have been introduced to take advantage of dimensional co-ordination, but the change is unpopular with bricklayers and failure to achieve one standard metric brick complicates and increases cost of brick production. Metric bricks are currently available in the following format sizes:

Length (mm)	Width (mm)	Height (mm)
300	× 100	× 100
200	× 100	× 100
300	× 100	× 75
200	× 100	× 75

Actual size are 10 mm less each dimension except the 300 mm length, which is 12 mm less.

Internal walls

The fundamental function of internal walls is to partition or divide the floor area of a dwelling into compartments or rooms. The materials employed are either concrete blocks or a timber framework with suitable cladding. This is known as a stud partition. Bricks are rarely used unless a high degree of fire resistance is required or for party wall construction between terraced and semi-detached dwellings, where sound insulation is an important criterion. A party wall is not classed as a partition, it is a wall which separates two adjacent properties and will be considered separately.

Partitions are classified as load bearing and non-load-bearing. Load-bearing partitions have to support their own dead load plus superimposed loading from a floor, roof or upper partition.

Non-load-bearing partitions support only their own weight and applied finishes, but they must be of sufficient strength to resist deformity from their own weight or lateral pressure.

Concrete block partitions

The Building Regulations require internal load-bearing partition walls of block construction to have lateral restraint by floor or ceiling joists at the top of each storey. The minimum thickness of internal load-bearing walls is half the thickness given in the table shown in Table 7.1 less 5 mm. For load-bearing partitions in the lowest storey of a three-storey building, the same regulation applies or 140 mm, whichever is greater.

As a guide to the use of concrete blocks for non-load-bearing partitions, it is recommended that the thickness should not be less than $\frac{1}{40}$ of the unrestrained height or length. The thickness may include 12.5 mm for cement and sand rendering on each face, bringing the total thickness of a 60 mm block wall up to 85 mm. Hence a maximum height or length of 3.40 m. However, practicalities often eliminate the use of 60 mm blocks, as their slender nature provides difficulty in laying more than a few courses at any one time. Supplementary support from vertical timber struts is needed, therefore 75 mm blocks capable of self-support are preferred.

Suitability of blocks

Concrete blocks are produced from Portland cement and aggregates selected from a wide range of sources. The quality of aggregates available for block production is extensive, and British Standard 6073 : 1981 reduces the classification to size and compressive strength. Sizes range from 390 to 590 mm long, 140 to 290 mm high and 60 to 250mm thick. Many contain voids, as shown in Fig. 7.15. For the exact

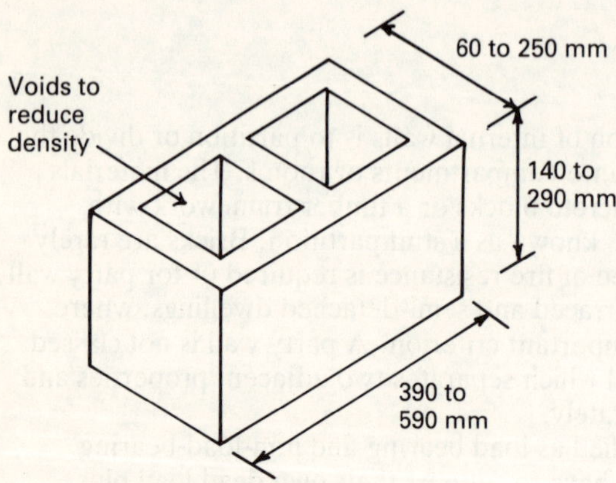

Fig. 7.15   Concrete block showing BS size range

size range refer to Part 2 of the British Standard. Compressive strength categories are 2.8, 3.5, 5, 7, 10, 15, 20 and 35 N/mm^2 and these result from the average of crushing 10 samples. The British Standard qualifies the lowest possible strength as 80 per cent of the test figure.

Methods of restraining block walls to prevent lateral movement at the ends are illustrated in Fig. 7.16. Where joists build into the wall, this is acceptable head restraint, but where joists run parallel with the partition, the technique illustrated in Fig. 7.17 could be used. Ground floor partitions are built up from a simple strip foundation or thickening of the ground floor slab as shown in Fig. 7.18. Upper floor partitions with no continuity from below require support as detailed in Fig. 7.19, the steel joist being employed for load bearing or heavy partitions.

Timber stud partitions

Timber stud partitions are rarely load bearing, they are therefore light in weight and supported by unmodified floor structure. The construction illustrated in Fig. 7.20 contains a simple framework of timber with head and sole plates secured to floor and ceiling respectively. The studding is spaced between 450 and 600 mm with noggings or struts to restrict movement. Openings for doors and

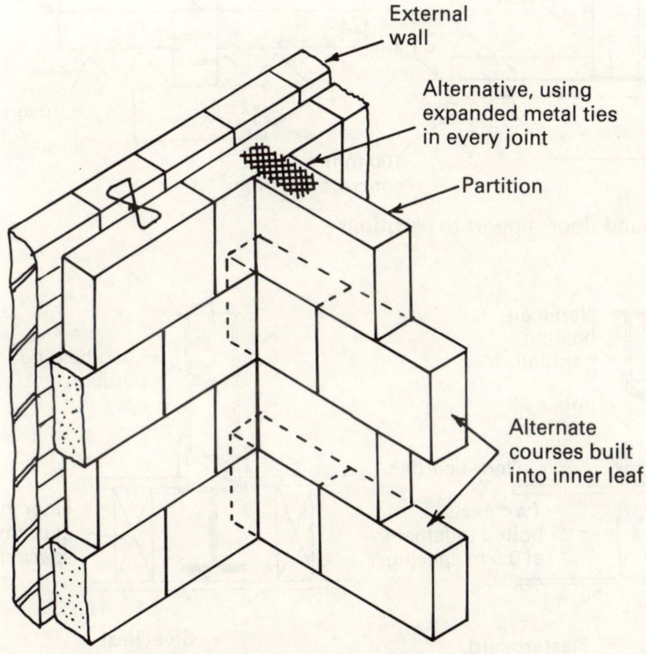

Fig. 7.16   Block bonding

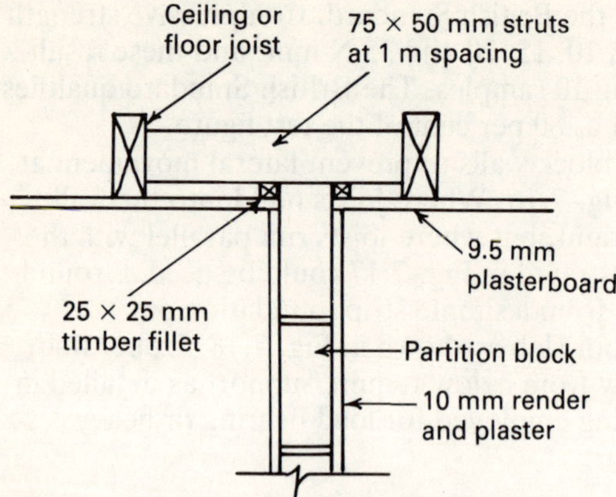

Fig. 7.17   Head restraint to block partitions

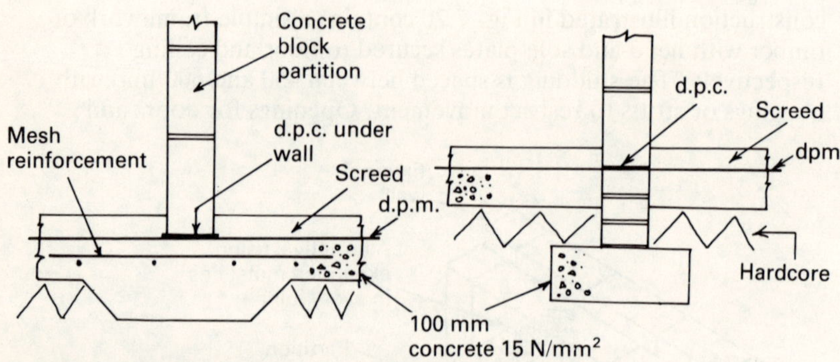

Fig. 7.18   Ground floor support to partitions

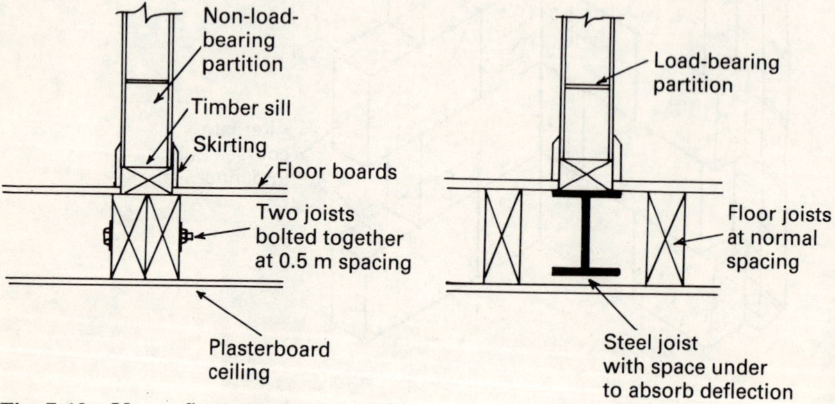

Fig. 7.19   Upper floor support to partitions

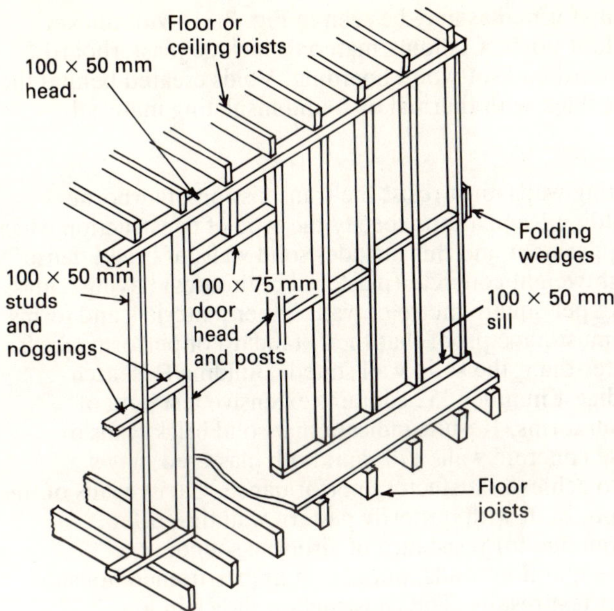

Fig. 7.20 Timber stud partition

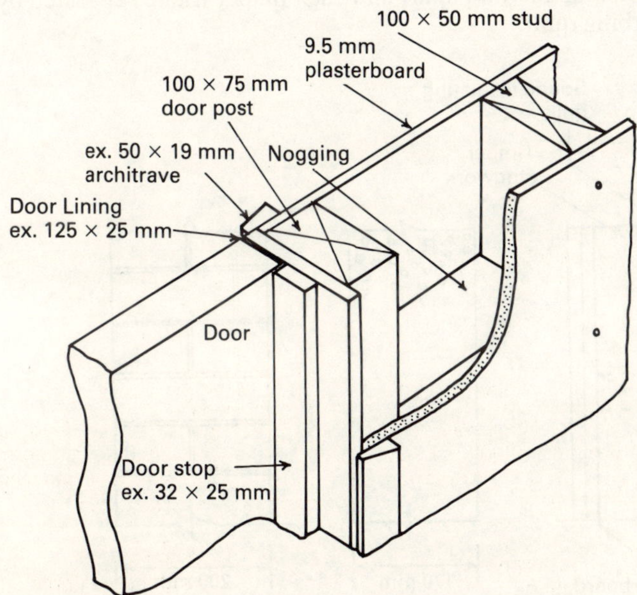

Fig. 7.21 Door opening detail in a timber stud partition

hatches are easily formed as may be seen in Fig. 7.21, with thicker timber for the door posts. Cladding materials include plasterboard, plywood, hardboard and softwood boarding. Voids created behind the cladding may be filled with thermal or sound insulating material.

Party walls

Party or separating walls must resist the transmission of airborne sound. The Building Regulations specify the type of construction which satisfies this requirement and this includes solid walls of brick, dense concrete and lightweight concrete (plastered both sides) having a mass of at least 415 kg per square metre of wall. Other materials and forms of construction must have proof that their standard of performance is equal to, or better than, the solid wall stated. Building Research Establishment digest number 252 contains extensive test data of various party wall forms. Results indicate that solid brick walls of 200 mm or dense concrete walls 170 mm, both plastered, have sufficient mass to achieve satisfactory performance. Cavity walls of the same composition, built with butterfly pattern wall ties, offer insignificant advantage for resistance of airborne sound.

Lightweight separating walls, popular in timber-framed housing, provide excellent test results. The construction shown in Fig. 7.22 compares favourably with brick and concrete walls, providing the wall is composed of two separate timber frames each externally faced with at least 32 mm of plasterboard. The inner faces of the plasterboard being a minimum of 200 mm apart and each timber frame separated by a sound-absorbing quilt.

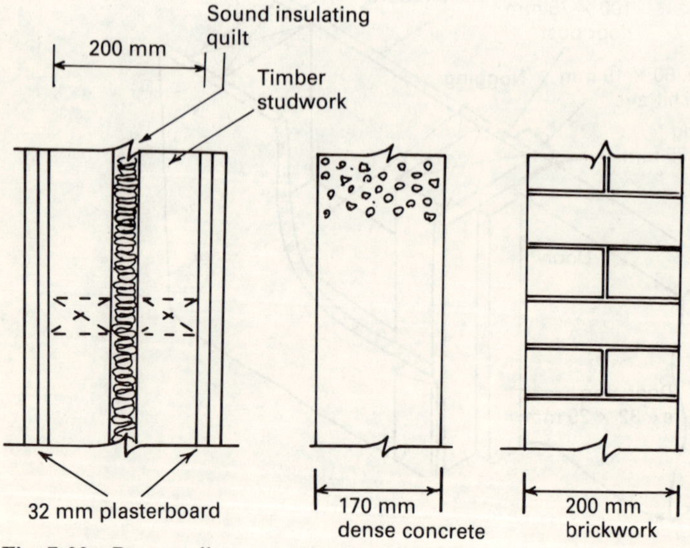

Fig. 7.22   Party wall construction

Chapter 8

Wall construction

Bricks and their manufacture

Bricks are the most popular of wall construction materials. The majority are produced from baked clay, others from autoclaved sand and lime and some from coloured concrete, but these are less widely used.

Clay bricks

The Romans were the first to introduce kiln-fired bricks to this country, but clay for brick manufacture was not really exploited until the Middle Ages. Until the Industrial Revolution and the introduction of the Hoffman kiln, brick manufacture was a combination of numerous manual processes. Bricks previously hand made in simple timber moulds were mass produced by machine and completely baked and fire in one section of the Hoffman kiln. Before modern firing techniques existed, bricks were processed in one kiln with fires ignited in grates in the outer walls. Alteratively, clamps were made by close formations of 'green' bricks having granules of coal mixed in with the clay. A fire was made from coal or rough breeze laid under the bricks and this spread to ignite the fuel contained in the bricks. The clamp burnt for several days and was then left to cool. In some brickworks this technique is still retained for the production of stock bricks.

The principle of the Hoffman kiln is shown in Fig. 8.1. It is a circular continuous tunnel containing 12 to 20 chambers with facilities for continuing the fire around the chambers. Thus, bricks in one

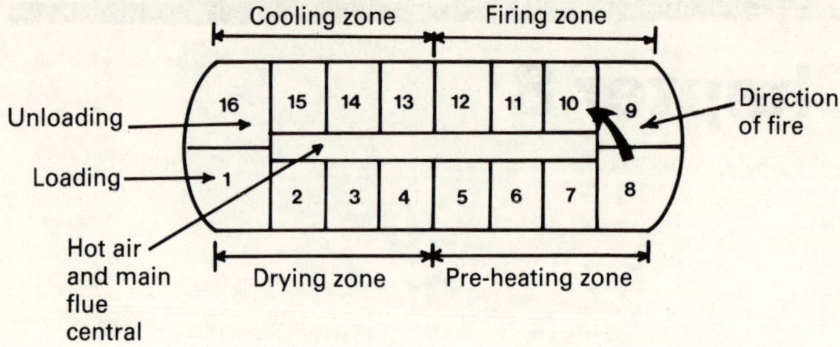

Fig. 8.1   Hoffmann kiln

chamber are dried, pre-heated, fired to about 1100 °C and cooled
without transporation.

A more recent innovation is the straight tunnel kiln, with central
firing zone. Here bricks are loaded on special wheeled carriers and
transported slowly through the increasing and then decreasing heat
zones. Variation in track speed provides specific functional or
appearance characteristics.

Calcium silicate (sand lime) brick

Calcium silicate bricks have a much shorter history than clay bricks.
Artificial stone production from steam treatment of caustic lime and
sand was first patented in this country in 1866. As a process it failed
because of the long production time associated with live steaming, but
in 1894 a German process using pressurised steam considerably
accelerated manufacture. These bricks are now very popular in
Germany and other countries where suitable deposits of clay for brick
production are scarce.

The manufacturing process commences with thorough mixing of
sand or crushed flint with lime, water and colouring if required. The
mix is then compressed and formed into brick units which are stable
enough to be stacked on carriers and wheeled into the autoclave.
Autoclaving is an extension of the principles used in domestic pressure
cooking with the bricks subject to compressed saturated steam for
several hours until the sand constituent has reacted with the lime and
water to form calcium hydrosilicates. The result is a very durable and
strong building brick which compares favourably in price with clay
bricks, but is less popular because of its uniformity of texture and
colour.

Bonding

Bonding is the arrangement or pattern of bricks in a wall. Before
considering some possibilities it is necessary to appreciate the features
of a brick and the variations and terms applied to cut bricks. Figure 8.2
illustrates brick dimensions and terminology and some examples of

bricks cut for featurework and use in bonding. The following bricklaying terms supplement these diagrams:

Stretcher – longest face of a brick.
Header – shortest face of a brick.
Half bat or snapheader – a brick cut midway along stretcher face.
Three quarter bat – a brick cut three quarters along the stretcher face.

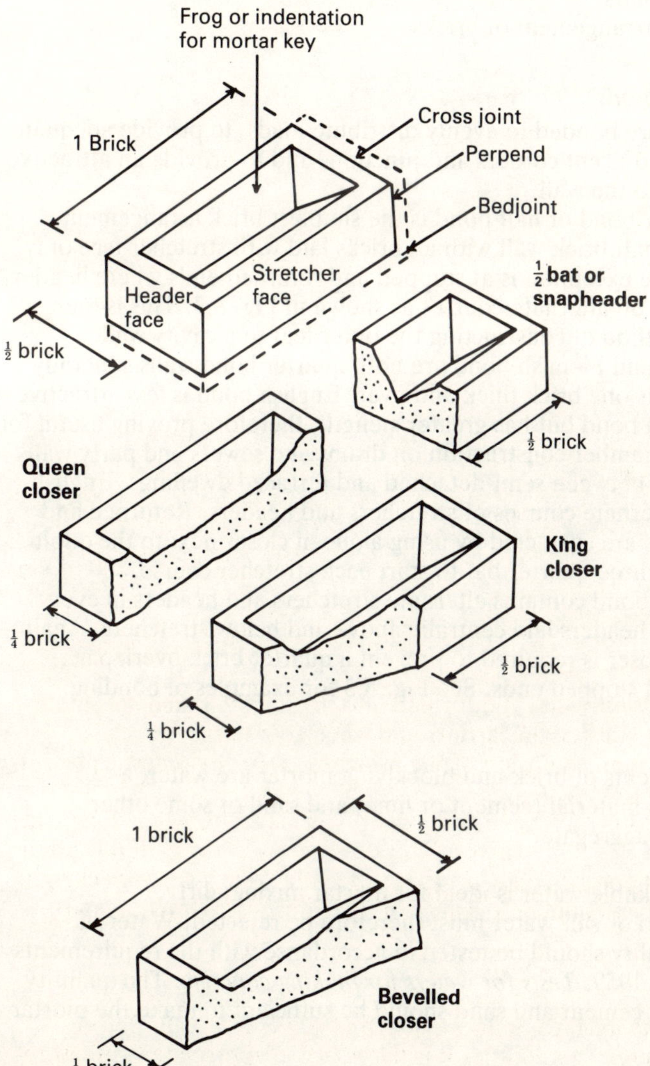

Fig. 8.2   Brick terminology

Queen closer – a brick cut centrally down its length.
King closer – a brick with bevelled cut removing half the stretcher face and half the header face.
Quoin – external angle of a wall.
Course – all bricks laid between two consecutive horizontal mortar joints.
Bed joints – mortar between two brick courses.
Cross joint or perpend – vertical joint, square to and connecting bed joints.
Bond – arrangement of bricks.

Bonding methods

Brick walls are bonded to evenly distribute loads, to provide adequate restraint to adjacent corners and junctions and to provide an attractive appearance to the wall.

Stretcher bond or half bond is the simplest brick arrangement providing a half brick wall with all bricks laid with stretcher face only exposed. The exception is at stopped and returned ends where header faces appear on alternate courses as shown in Fig. 8.3. This is the standard method of constructing the outer leaf to a cavity wall.

English and Flemish bond are both quarter bond walls generally used for walls one brick thick and over. English bond is less attractive than Flemish bond but has greater stength, therefore proving useful for inspection chamber construction on drains and sewers and party wall construction between semi-detached and terraced dwellings. English bond has alternate courses of stretchers and headers. Returned and stopped ends are corrected by using a queen closer next to the quoin header or a three quarter bat to start each stretcher course.

Flemish bond contains alternate stretchers and headers in every course, with headers laid centrally above and below stretchers. Again the queen closer is required to prevent a quarter brick overlap at returned and stopped ends. See Fig. 8.3 for examples of bonding.

Mortar

The components of brick and blocklying mortar are water, a cementitious material (cement or lime) and sand or some other suitable fine aggregate.

Water. Drinkable water is ideal for mortar mixing; dirty, contaminated or salt water must therefore be rejected. Water of doubtful quality should be tested in accordance with the requirements of BS 3148 : 1959: *Tests for waters for making concrete*. The quantity added to the cement and sand should be sufficient to make the mortar workable.

Cement. The two cements normally used are ordinary Portland cement and Portland blast furnace cement. Rapid-hardening Portland cement

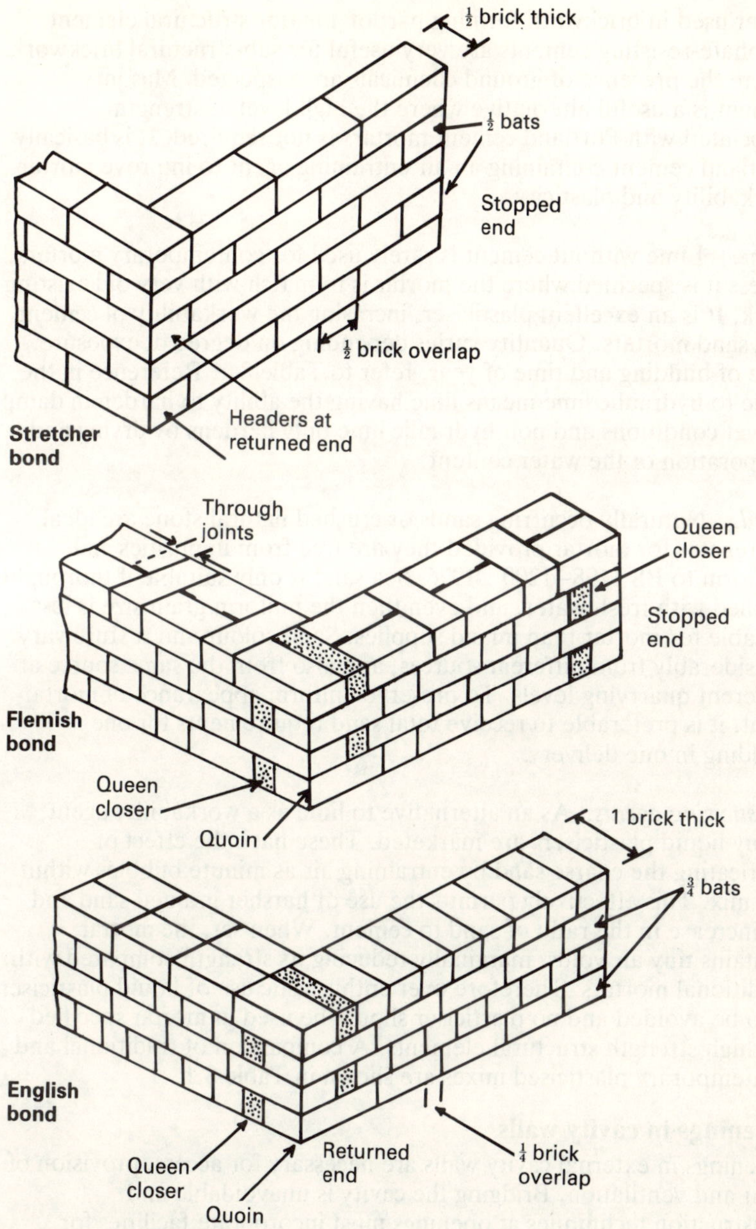

Fig. 8.3 Brick bonding

may be used where a fast set is required and high alumina cement could be used to ensure high early strength. High alumina cement has sensitive properties and should only be used under strict control and

110

never used in brickwork forming part of a major structural element. Sulphate-resisting cements are very useful for sub-structural brickwork where the presence of ground chemicals are suspected. Masonry cement is a useful alternative where the high level of strength associated with Portland cement mortars is not required. It is basically Portland cement containing an air entraining agent to improve mortar workability and plasticity.

Lime.   Lime without cement is rarely used for contemporary mortars, unless it is specified where the mortar is to match with very old existing work. It is an excellent plasticiser, incrasing the workability of cement and sand mortars. Quantity varies depending on degree of exposure, type of building and time of year, refer to Table 8.1. Reference in the table to hydraulic lime means lime having the ability to harden in damp or wet conditions and non-hydraulic lime only hardens by drying and evaporation of the water content.

Sand.   Naturally occurring sands or crushed natural stone are ideal aggregates for mortar provided they are free from impurities and conform to BS 1198–1200 : 1976. Sea sand is only suitable if thoroughly washed with fresh water, and even then the uniform grain size is less suitable for mortar than inland supplies. Sand colour and texture vary considerably from different sources, and also from the same source at different quarrying levels. To preserve uniform appearance of mortar joint, it is preferable to receive total sand requirements for one building in one delivery.

Plasticising agents.   As an alternative to lime as a workability agent, many liquid plasticisers are marketed. These have the effect of lubricating the coarse sand by entraining air as minute bubbles within the mix. This effectively permits the use of harsher grained sand and an increase in the ratio of sand to cement. When set, the mortar contains tiny air voids, marginally reducing its strength compared with traditional mortars. Therefore over enthusiastic use of liquid plasticiser is to be avoided and no plasticiser should be used in mortar specified for high strength structural elements. A comparison of traditional and contemporary plasticised mixes are shown in Table 8.1.

Openings in cavity walls

Openings in external cavity walls are necessary for access, provision of light and ventilation. Bridging the cavity is unavoidable and construction techniques at openings must incorporate facilities for preventing moisture crossing the void. The construction details in Figs. 8.5 to 8.11 each include a damp proof courses or have dampness barriers included with their manufacturer. Sills are moulded with weathered edges, anti-capillarity grooves and drips. Lintels are stepped forward or provided with a stepped damp proof course to divert dampness.

Table 8.1 Mortar mixes

	Type of construction	Exposure	Time of year	Traditional mix	Masonry cement and sand	Cement and sand with plasticiser
External walls. Clay, concrete and sand lime bricks	Above dpc	Sheltered	Spring and summer	1 : 2: : 8 cement, lime and sand 1 : hydraulic lime and sand	1 : 6	1 : 7
		All conditions	Autumn and winter	1 : 1 : 5 cement, lime and sand. 1 : 2 hydraulic lime and sand	1 : $4\frac{1}{2}$	1 : 5
	Below dpc, parapets and free standing walls.	All conditions	All seasons	1 : 1 : 5 cement, lime and sand 1 : 3 cement and sand		
Engineering bricks	Engineering construction and work below ground	All conditions	All seasons	1 : 3 cement and sand		
Internal walls			Spring and summer	1 : 3 : 10 cement, lime and sand 1 : 3 hydraulic lime and sand	1 : 7	1 : 8
			Autumn and winter	1 : 2 : 8 cement, lime and sand 1 : 2 hydraulic lime & sand	1 : 7	1 : 8

The threshold is the lowest opening in an external wall. The main feature is the hardwood sill mounted above the damp proof course. It has a sloping or weathered edge and drip to the underside. Unlike a window sill, it has a galvanished steel or plastic water bar recessed into the upper surface to prevent driven rain and draughts penetrating beneath the door. The possible construction at threshold is shown in Fig. 8.4. This includes an internally opening door complete with weather moulding designed to deflect rain running down the door face onto the weathered surface of the sill.

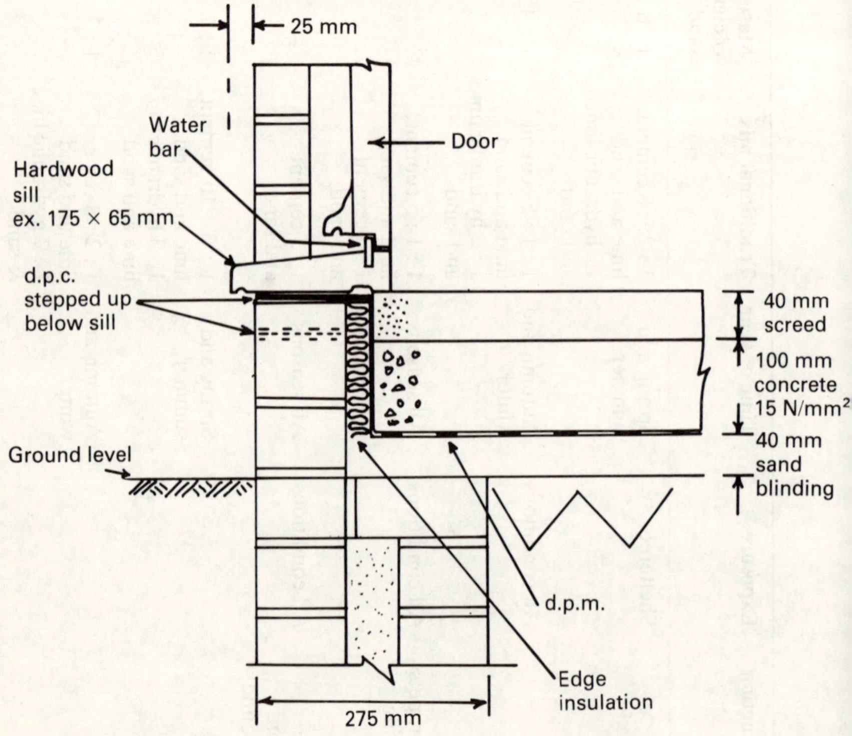

Fig. 8.4 Threshold construction

Window sill treatment offers numerous construction alternatives, with the possibility of attractive features created by tile or brick subsills. The standard sill and associated wall construction are shown in Fig. 8.5 with an alternative feature window sill incorporating tiled subsill. Notice the reduced sill where a subsill is provided and the internal rebate is in both sills for location of window board.

Both door and window jambs are treated similarly where they are secured to the wall construction. Here it is necessary to close the cavity by returning the inner leaf with cut blocks or purpose-made blocks. To

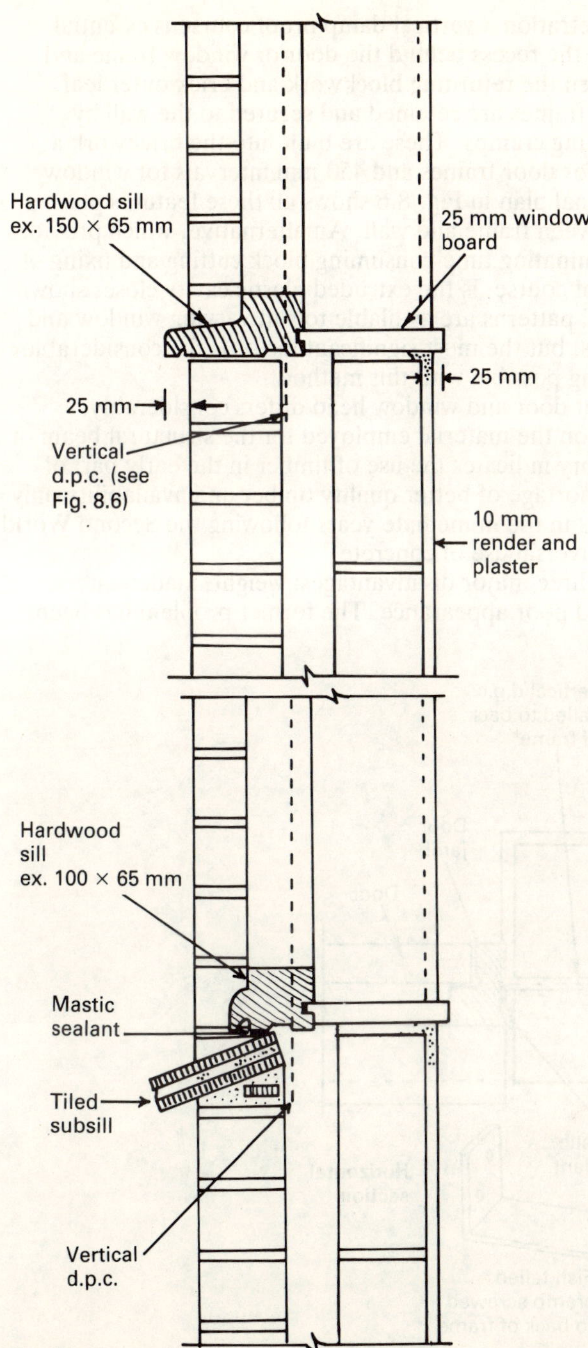

Fig. 8.5 Construction at window sill

prevent water penetration a vertical damp proof course is essential. This is tacked into the recess behind the door or window frame and sandwiched between the returning blockwork and brick outer leaf. Door and window frames are retained and secured to the wall by galvanised steel fixing cramps. These are built into the brickwork at 300 mm intervals for door frames and 450 mm intervals for window frames. The sectional plan in Fig. 8.6 shows all these features plus mastic sealant between frame and wall. An alternative, which provides a frame fixing, eliminating time-consuming block cutting and fixing of vertical damp proof course, is the extruded plastic cavity closer shown in Fig. 8.7. Several patterns are available to suit varying window and door frame designs, but the most significant factor is the considerable cost and time saving possible with this method.

Construction at door and window head differs considerably, depending mainly on the material employed for the structural beam or lintel. Recent history indicates the use of timber in the early part of this century, but shortage of better quality timber and availability only of licensed supplies in the immediate years following the Second World War, led to the universal use of concrete.

Concrete has three major disadvantages: weight, inadequate tensile strength and poor appearance. The former problem has been

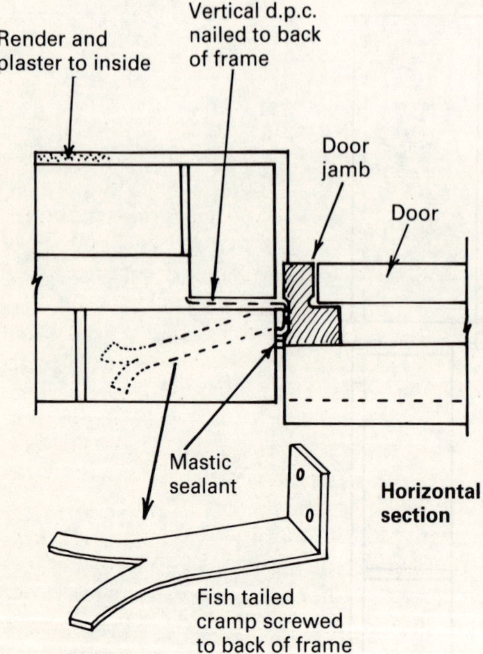

Fig. 8.6 Construction at jamb

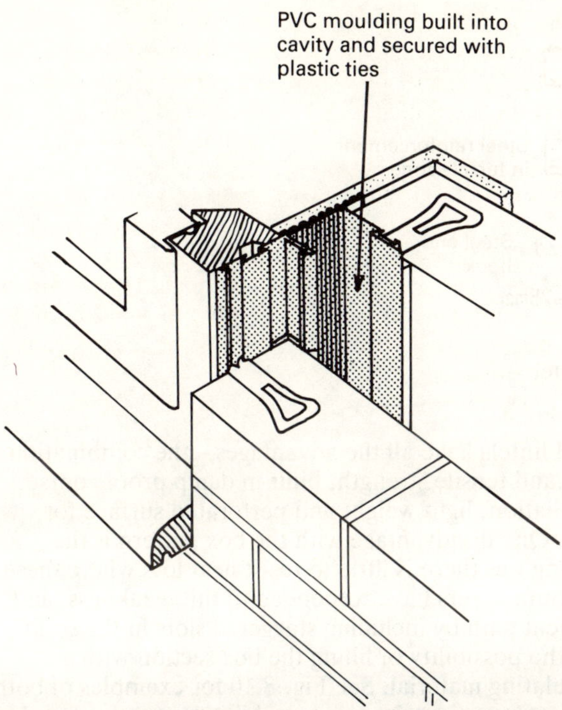

Fig. 8.7 'Dacatie' PVC cavity closer and jamb fixing

partially relieved by pre-stressing, which reduces the concrete content. Tensile strength is provided by steel reinforcement, but the appearance is only relieved by providing a steel angle to support external brickwork. Production of a composite unit (steel-reinforced concrete) or installation with a steel angle complicates construction, hence the attraction of lightweight pressed galvanised steel box or plain section lintels, which now virtually dominate the lintel market.

However, many openings with reinforced concrete lintels do exist; therefore before considering the steel lintel the principles of concrete lintels will be considered. Details in Fig. 8.8, indicate the effect of loading a concrete lintel, the lower portion of which is subject to tensile stresses. Here, steel reinforcement is essential. A minimum cover of 25 mm concrete is required to provide sufficient bonding to the steel, in addition to protection from corrosion and fire. End hooks are desirable to provide secure bondage and the recommended bend profile is shown. Two construction details are illustrated in Fig. 8.9, one with galvanised steel angle and soldier arch to hide the concrete lintel and the other with a boot lintel, designed to reduce the amount of exposed concrete. Notice the need for a stepped damp proof course in each case.

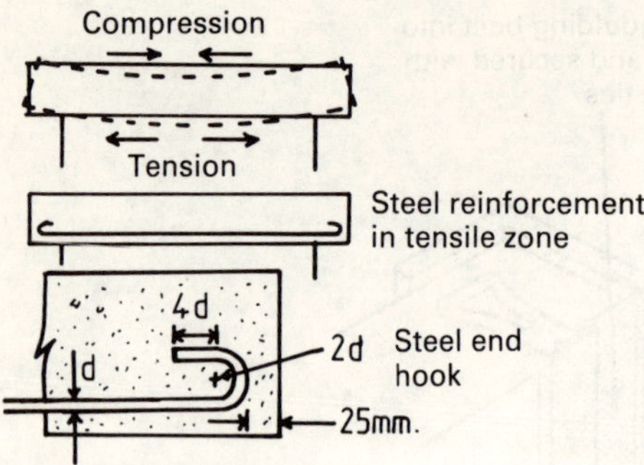

Fig. 8.8 Concrete lintel

Galvanised steel lintels have all the advantages – the combination of high compressive and tensile strength, built-in damp proof course, quick and easy installation, light weight and perforated surface for efficient plaster key. One disadvantage with the box pattern is the effect of 'cold bridging', as there is little to resist heat loss where these lintels are used. Manufacturers have responded to this weakness, and have extended the heat path by including staggered slots in the soffit plate. There is also the possibility of filling the box section with a suitable thermal insulating material. See Fig. 8.10 for examples of both the pressed steel fore-runner in galvanised steel lintels and the popular box section pattern.

Arches

Arch construction is time-consuming, skilled work, two factors rationalised by most modern building techniques. However, recent exploitation of brick feature work and the demand for housing with traditional features has inspired the reintroduction of arched openings. Figure 8.11 combines two semi-circular arches, one with wedge-shaped stones or voussoirs, the other with tapered or gauged brickwork.

Rough brick arches are only suitable for flat or slightly cambered arches. These are uncut bricks which leave a large wedge-shaped joint on curved arches. Axed bricks are preferred in this situation and these are bricks cut to a taper, thus providing a mechanically balanced structure with neat parallel joints. Gauged bricks are the most attractive and accurate means of forming an arch. They are usually 'softer' than the other facing bricks, possessing ideal cutting and finishing properties. Some gauged arches contain very precisely cut and ground bricks, which permit a fine but distinctive joint between each brick.

Arch shapes may resemble many geometric forms including semi-circular, segmental, semi-elliptical, mitred or pointed. Temporary

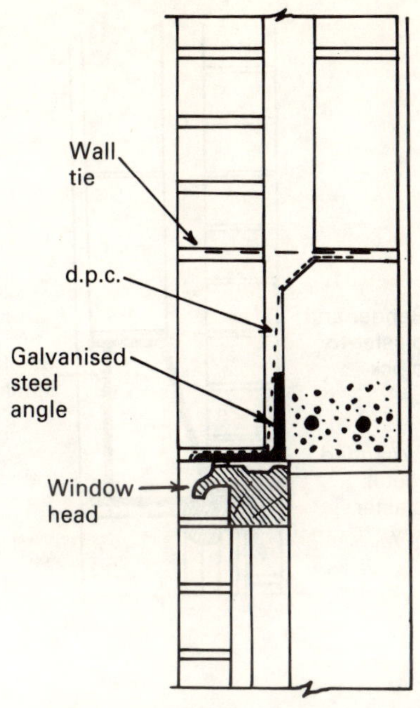

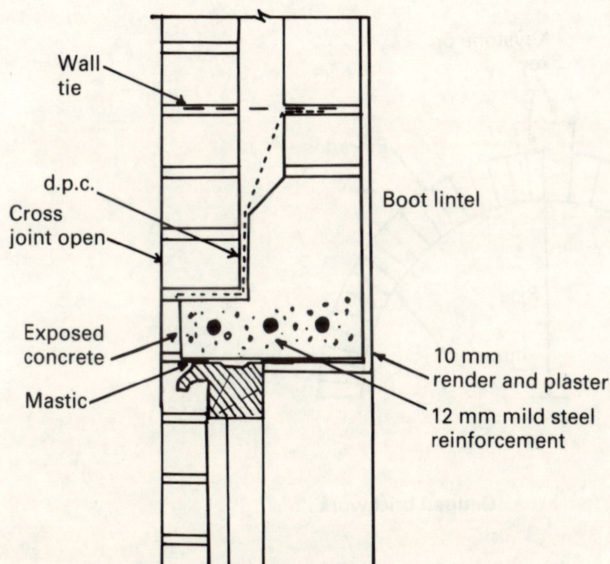

Fig. 8.9   Vertical sections of reinforced concrete lintels

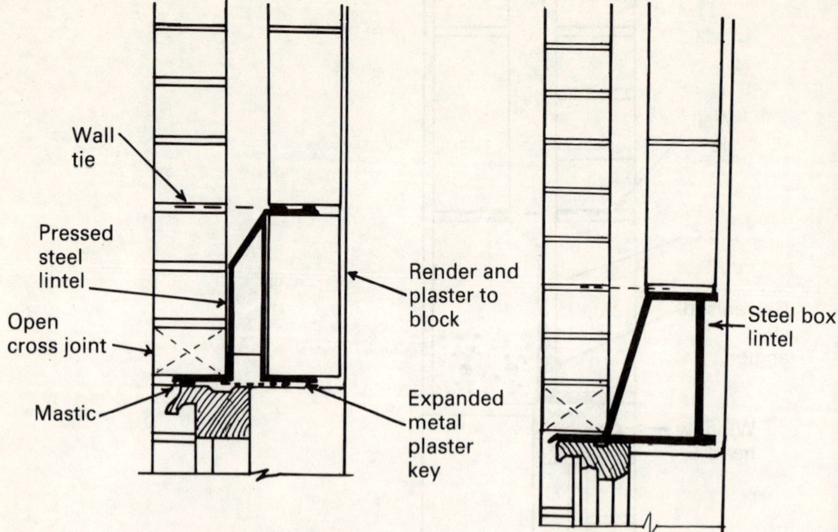

Fig. 8.10   Galvanised steel lintels

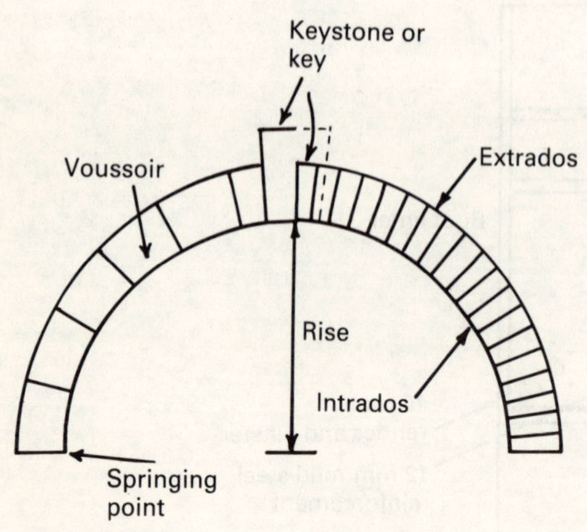

Fig. 8.11   Semi-circular arch

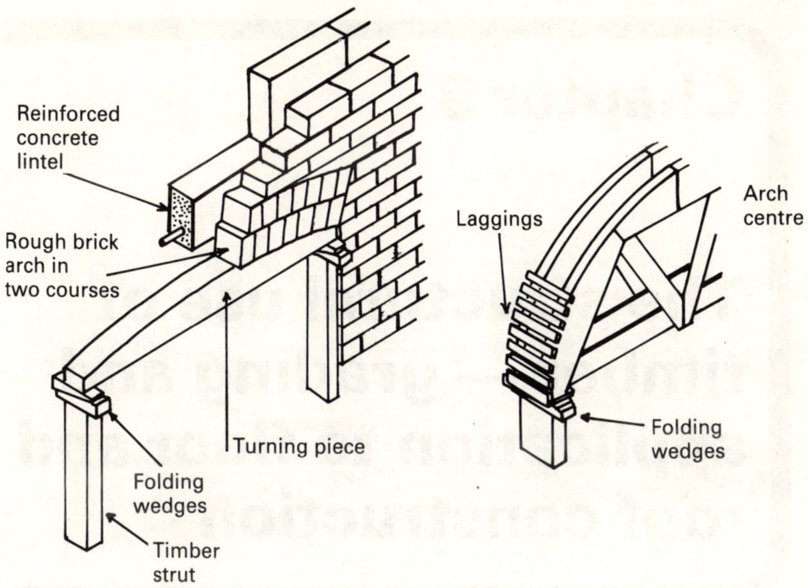

Fig. 8.12 Temporary support during arch construction

support is provided while the arch is formed until the mortar has gained sufficient strength. This support is known as a centre, or for very low rise arches, a turning piece. They are produced from timber and should follow the exact line of the arch curve. Laggings or short cross strips of timber are provided for direct bearing of the arch bricks. Folding wedges are provided to each side at the top of timber props; these ease removal of the centre without damage to the wall. See Fig. 8.12 for examples of a timber centre and the use of a turning piece.

Chapter 9

The structural use of timber — grading and application to floor and roof construction

Structural grading of timber

The demand for timber in building and other industries has imposed excessively upon the world's resources. Timber grown in Britain supplies only a small amount of the nation's requirements, therefore, we rely considerably on exports from North and South America, Scandinavia, Russia and Africa. Increasing demand has encouraged rapid growing techniques and trees rarely achieve maturity before felling. Structural softwood is all sapwood and defects due to knots, shakes and waney edge predominate. The extent of defects vary from species to species, and even two pieces of timber from the same log can have different strength characteristics. The use of BS 4978 : 1973 provides a reliable method of comparing and grading structural timber by sorting sawn sections into categories of predictable strength. This permits reliable use of design data for calculation of safe sections, spans and spacings for joists and beams.

Grading of timber may be visual or by the more efficient use of machinery. It may be graded in this country or the country of origin, but in the absence of international grading standards, confusion between BS 4978 gradings and foreign suppliers' grades is possible. Tables 9.1 and 9.2 show the grading provided by the British Standard and the Canadian National Lumber Grades Authority, respectively. Grading is in descending order within the groupings, but for convenience only, are grouped A to F. Within the BS grades, MGS is deemed comparable to GS, and MSS to SS. Exact comparison of BS

Table 9.1 BS 4978 Grades

Mark	Grade
	A. Visual grades
SS	Special structural
GS	General structural
	B. Machine grades
M75	Machine 75
MSS	Machine special structural
M50	Machine 50
MGS	Machine general structural

grades and NLGA grades is less simple, but NLGA No. 2 of hem-fir or spruce-pine-fir species groups will satisfy GS or MGS grade specifications for redwood/whitewood.

For easy identification, both the BS and Canadian grading rules require markings on graded timber. Various markings exist, and two possibilities are shown in Fig. 9.1, the first for visual graded timber and the other for machine-graded timber. Some of the more common species of timber available in BS 4978 grades include European red and whitewood and British-grown scots pine, douglas fir and sitka spruce. Imported NLGA graded timber includes hem-fir, douglas fir-larch and spruce-pine-fir.

Table 9.2 NLGA Grades

Mark	Grade	Range of nominal widths
	C. Structural light framing	50–100 mm
Sel str	Select structural	
No. 1	Number one	
No. 2	Number two	
No. 3	Number three	
	D. Light framing	50–100 mm
Const	Construction	
Std	Standard	
Util	Utility	
Stud	E. Stud	50–100 mm
	F. Structural joists and planks	150 mm and over
Sel str	Select structural	
No. 1	Number One	
No. 2	Number two	
No. 3	Number three	

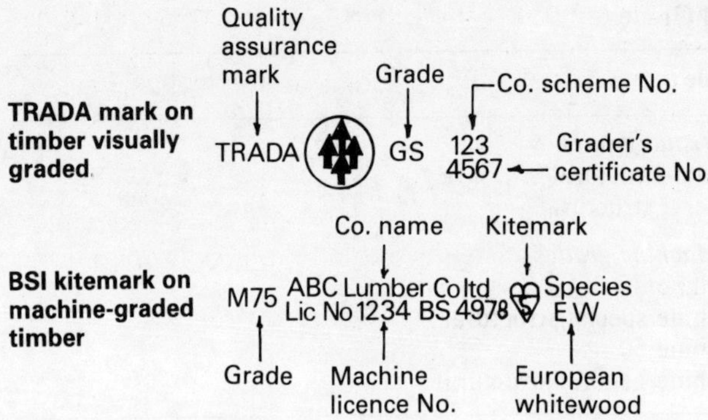

Fig. 9.1   Grade markings on timber

Characteristics for visually determining GS and SS grade timber

Knots, margin conditions and longitudinal separation.

The effect of knots is assessed by comparing the sum of the projected cross-sectional areas of knots with the cross-sectional area of the piece of timber. This is known as the knot area ratio and it may be affected

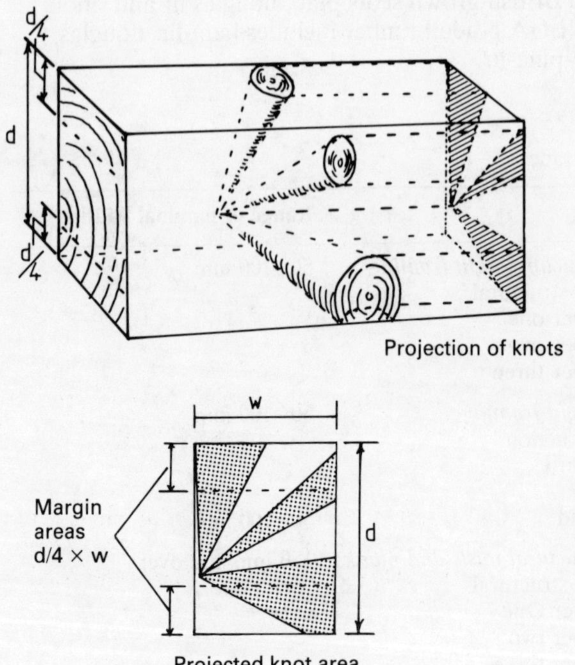

Fig. 9.2   Knot projection and margin areas

by margin conditions. Margins are determined as the areas adjoining the edge of cross-sections and each occupy one-quarter of the total section area, and a margin condition exists where over half the area of either margin contains the projected knot area. This is simply represented in Fig. 9.2.

Longitudinal separation refers to two or more knots or groups of knots, both having knot area ratios over 90 per cent of the permitted ratio and being separated in the lengthwise direction. For separation distances, see the table of grading characteristics.

Fissures and resin pockets

Fissures refer to separation of the annual growth rings, more usually known as shakes. Examples are shown in Fig. 9.3 and their size is determined by measuring the distance between lines projected from the fissure ends parallel to the opposite faces. Where fissures occur on opposite faces, and a line square to the section face cuts both, the size of fissure is taken as the sum of both. Resin pockets are measured in the same way.

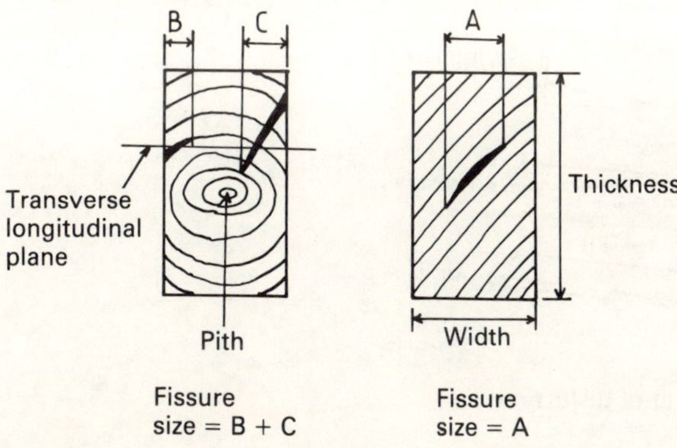

Fig. 9.3   Size of fissures

Slope of grain

Slope of grain is determined by scoring a line along the grain of the timber and comparing this with the parallel sides of the section.

Wane

Wane or waney edge refers to timber cut close to the outer surface of the log, having one or two incomplete corners. The amount of wane is measured parallel to the edge or face affected and expressed as a fraction of the surface dimension as shown in Fig. 9.4.

Growth rate

Growth rings are measured by averaging the number per 25 mm,

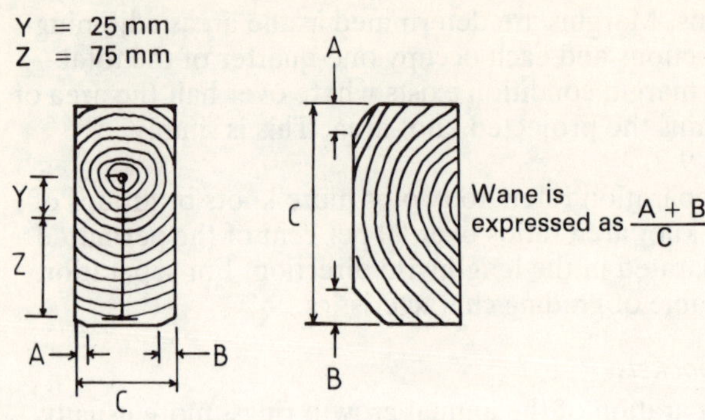

Fig. 9.4   Amount of wane and growth rate

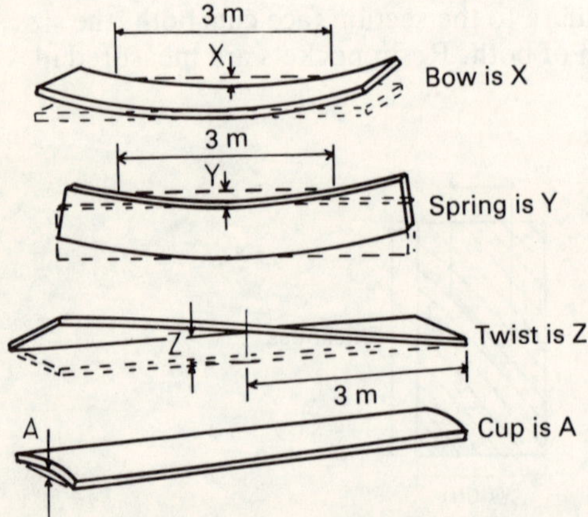

Fig. 9.5   Measurement of distortion

from a radial line 75 mm long. If the pith is present, the line should commence 25 mm beyond, and where 75 mm is impractical the longest possible line is taken. See Fig. 9.4.

Distortion

Distortion is measured over the length and width, and means of determining the amount are shown in Fig. 9.5. Table 9.3 shows the comparison between the characteristics and tolerances determining GS and SS grade timber.

Suspended timber ground-floor construction

Suspended timber ground floors were always considered the most satisfactory method of providing a quality floor structure. Since the

Table 9.3

Grading Characteristics	Grades	
	G.S.	**S.S.**
Knot area ratio	Maximum 1 : 2	Maximum 1 : 3
	If margin condition exists, maximum 1 : 3	If margin condition exists, maximum 1 : 5
	If cross-section is square, maximum 1 : 3	If cross-section is square, maximum 1 : 5
Longitudinal separation	Separation distance minimum $\frac{1}{2}$ timber width	Separation distance minimum $\frac{1}{2}$ timber face
Fissures and resin pockets	Defects less than or equal to $\frac{1}{2}$ timber thickness	Defects less than or equal to $\frac{1}{2}$ timber thickness
	Defects greater than $\frac{1}{2}$ timber thickness but less than thickness, fissures length maximum 900 mm or $\frac{1}{4}$ length of timber, take least value	Defects greater than $\frac{1}{2}$ timber thickness but less than thickness, fissures length maximum 600 mm or $\frac{1}{4}$ length of timber, take least value
	Defects equal timber thickness, fissures length maximum 600 mm. If fissures occur at end of the timber, their length is maximum $1\frac{1}{2}$ times width of timber	Defects equal timber thickness, fissures only permitted if they occur at end of timber, and their length is a maximum of the timber width
Slope of grain	Maximum 1 in 6	Maximum 1 in 10
Wane	Maximum $\frac{1}{3}$ of the surface dimension on which it occurs	Maximum $\frac{1}{4}$ of the surface dimension on which it occurs
	Within 300 mm of either end of timber, maximum $\frac{1}{2}$ surface dimension within single continuous length, maximum 300 mm	

Table 9.3 *Contd*

Grading Characteristics	Grades	
	G.S.	S.S.
Rate of growth	Minimum of 4 annual rings/25 mm	Minimum of 4 annual rings/25 mm
Distortion	Bow, maximum $\frac{1}{2}$ thickness/3 m length Spring, maximum 15 mm/3 m length Twist, maximum 1 mm/25 mm of width in any 3 m length Cup, maximum 1/25 of width.	

Note. Tolerances for distortion are not mandatory, but provided for guidance only.

Second World War, problems of timber shortage and escalating prices have reduced its popularity in favour of solid concrete floors. Labour costs too have added to the decline of timber floors, as this form of construction is time-consuming, skilled work. Nevertheless, suspended floors exist, and many are constructed to match existing structures when houses are extended. They are also provided to new houses where space for warm air central heating ducting canot be concealed in a solid floor, or in houses built on sloping sites as indicated in Fig. 2.5 (p. 14). Here the floor joists are suspended and supported by opposing walls.

Conventional suspended ground floors vary little from the detail in Fig. 9.6. This includes the requirements of Building Regulations C3 and C4, which were considered in part when referring to solid concrete ground floors in Chapter 6. The remaining part of Regulation C4 requires the upper surface of the oversite concrete to be at least the highest level of adjoining ground and a sound bed of sulphate-free hardcore must be provided beneath the concrete. Furthermore, sufficient air circulation must be provided beneath the floor surface by spaces of at least 75 mm from the upper surface of concrete and the underside of joists. Provision of access for air circulation beneath the timber floor is not defined in the Bulding Regulations. Building Research Establishment Digest No. 18 suggests a minimum of 3000 mm^2 open area in air bricks for every linear metre of external wall. The National House Builders Council rules are more practical, requiring airbricks at intervals of 2 m centres and no more than 450 mm from the end of the walls.

Sleeper walls are provided at 1.5 to 2 m spacing for intermediate support to floor joists. These have bricks omitted at regular intervals to provide a free passage of air under the floor. This is essential to prevent condensation and subsequent decay from wet or dry rot.

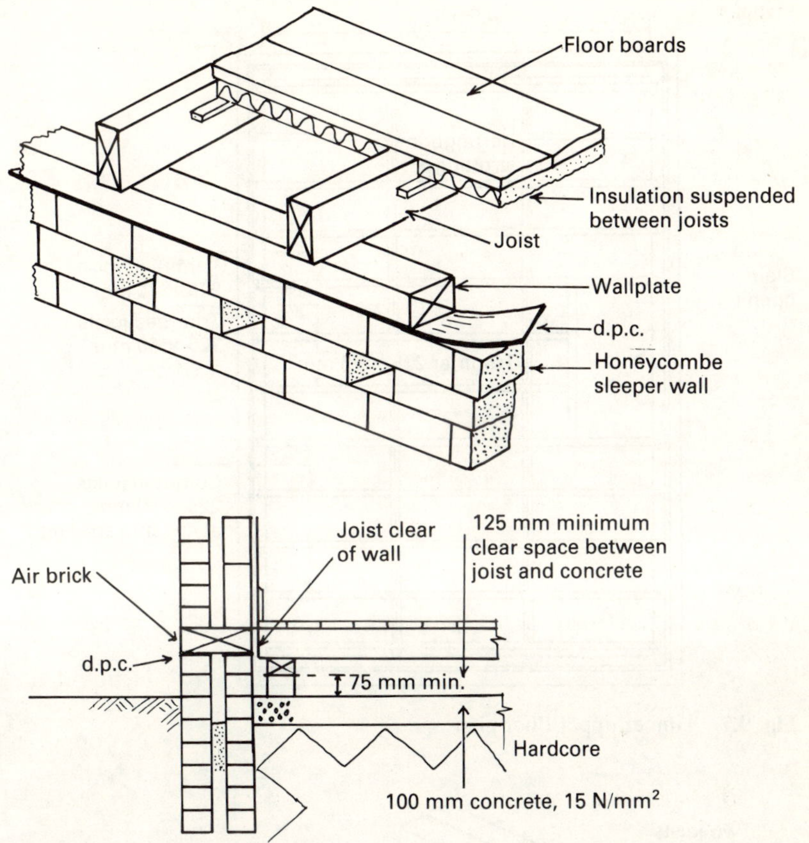

Fig. 9.6 Suspended timber ground floor

Timber upper floor construction

Timber upper floors used in house construction are classified as single or double. Single floors contain common joists which span from wall to wall and double floors have intermediate support to joists, provided by steel or timber beam.

The floor plan in Fig. 9.7 shows a typical upper floor constructed on single floor principles, but if joists are to span beyond 4.5 m it often proves more economical to provide a double floor. The construction at intersection of joists and beam is shown in Fig. 9.8, with galvanised steel joist hanger locating timber sections. In both examples the joists are notched to preserve uniform floor and ceiling levels.

End support to joists is from the cavity wall, and different methods of support are illustrated in Fig. 9.9. The wallplate provides a uniform level, a suitable joist fixing and spreads the floor load evenly, but as an

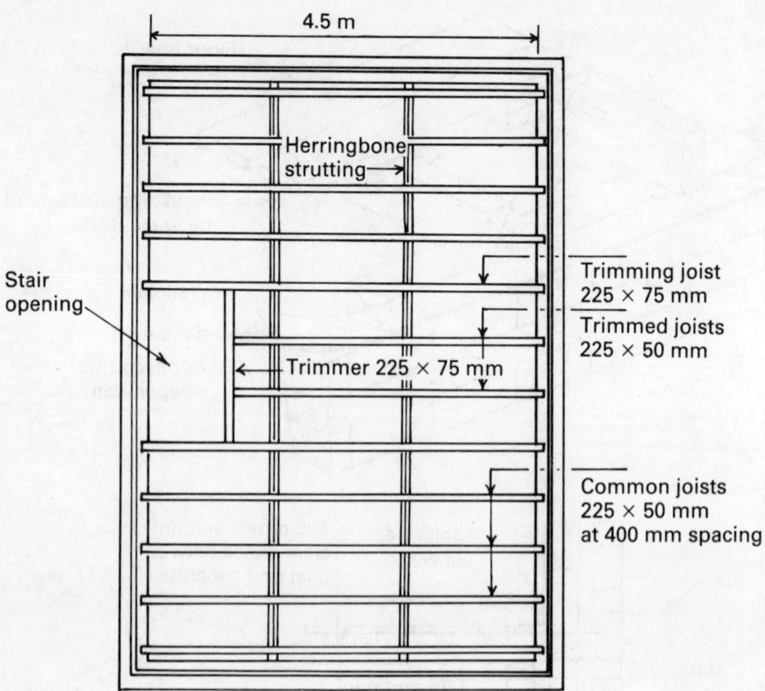

Fig. 9.7 Timber upper floor plan

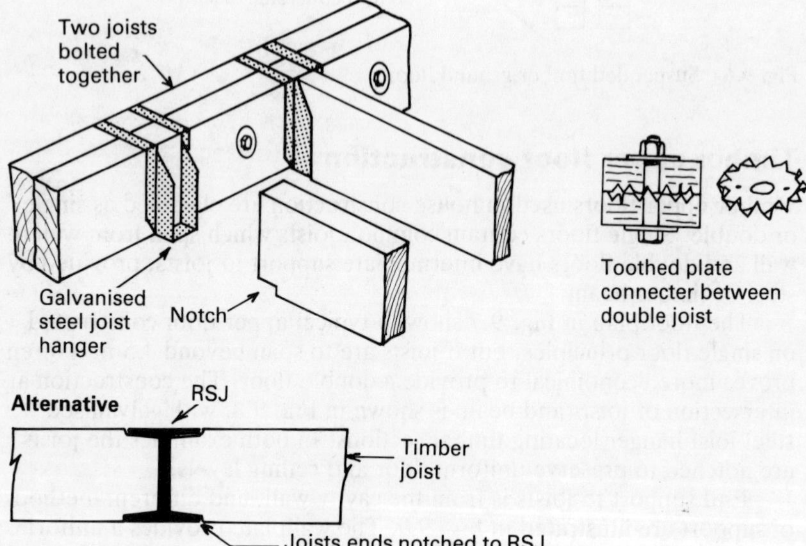

Fig. 9.8 Double floor construction

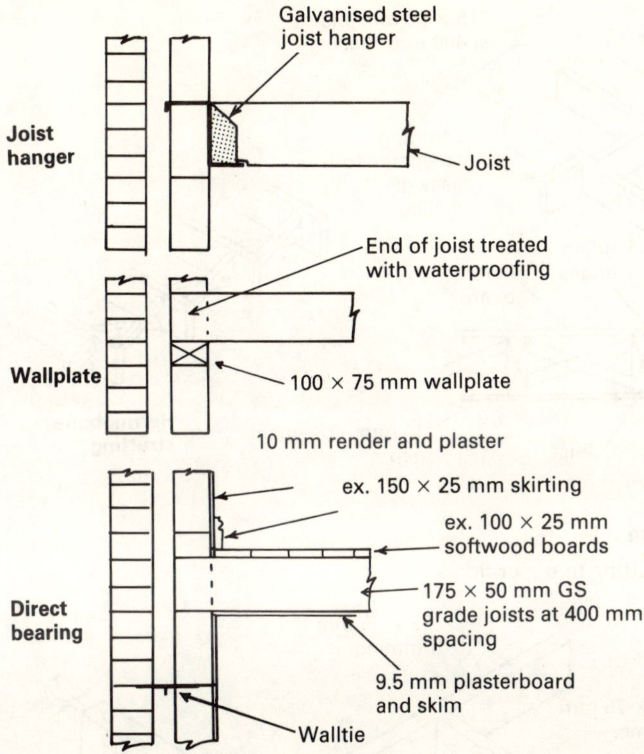

Fig. 9.9 Joist support at cavity wall

economy measure direct bearing on the inner leaf is now widely practised. Ends of joists must bear at least 90 mm (75 mm on a wallplate) and should not penetrate the cavity. A clear waterproofing treatment to the joist ends is advisable. Joist hangers provide a simple means of support without disturbing the wall bond, but care must be taken to ensure they are in alignment (see Fig. 9.9).

Stiffening and resistance to movement of the joists is provided by installing herringbone or solid strutting across the centre of the floor or at about 1.5 m spacing for larger spans. Herringbone strutting is produced by nailing struts of about 38 mm square timber between joists and arranged to cross from the top and bottom edges of adjacent joists in a continuous line. Tightening is by folding wedges between the first and last joists and the wall. Solid strutting is a simpler alternative, although less effective for restraining warping and twisting of the joists. Both methods are shown in Fig. 9.10.

The cutting and arrangement of joists around stair openings is formed with trimmer, trimming and trimmed joists as shown in

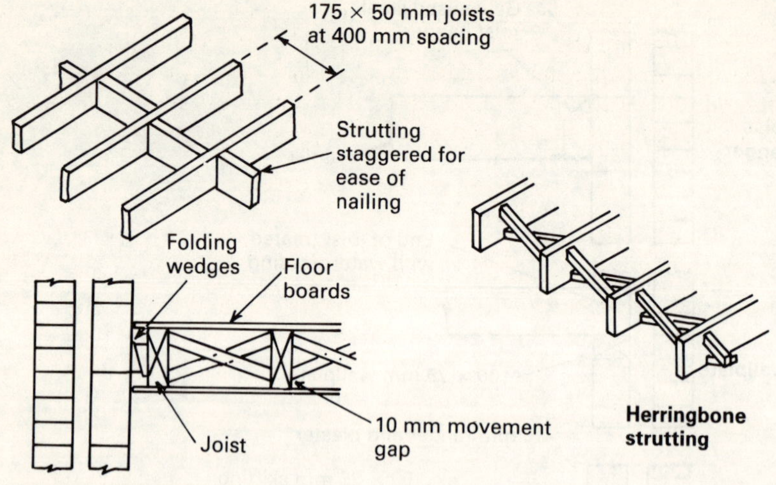

Fig. 9.10 Strutting to upper floors

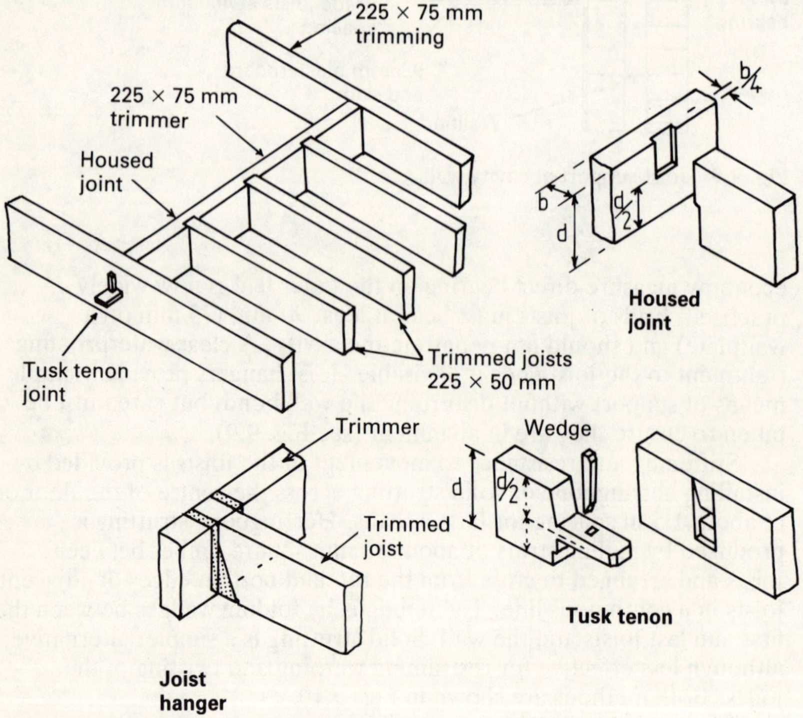

Fig. 9.11 Treatment at stair openings

Fig. 9.10. The trimmer and trimming joists are expected to carry greater loads than the common joists and under normal loading an increase of at least 3 mm in width over common joists for every trimmed joist carried will be sufficient. The details in Fig. 9.11 provide examples of traditional and modern jointing techniques.

Roof construction

Roof form depends on the slope or pitch and the span between supporting walls. Construction is by combining several structural components, most of which indicated in Fig. 9.12.

Traditional roofs are classified for convenience as single, double and triple, and modern pre-fabricated roof trusses are known as trussed rafters. Single roofs are the simplest roof form, containing rafters secured to ridge board and wallplate. These include flat (maximum pitch, 10°), lean-to, couple, close couple and collar roofs shown in Fig. 9.13. Spans are limited to about 4.5 m using realistic sizes of timber; for greater spans additional structural members are required.

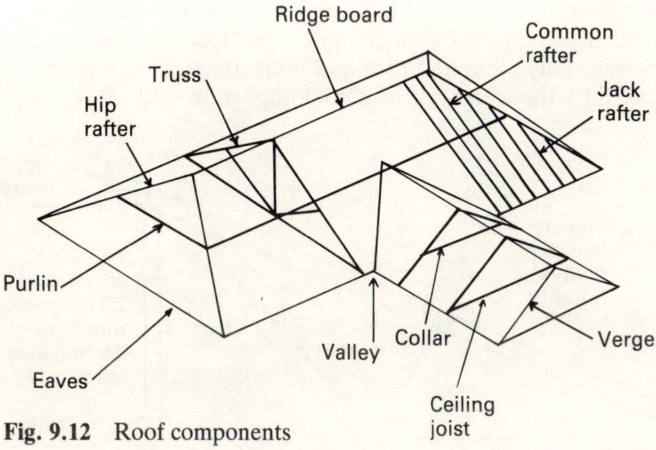

Fig. 9.12   Roof components

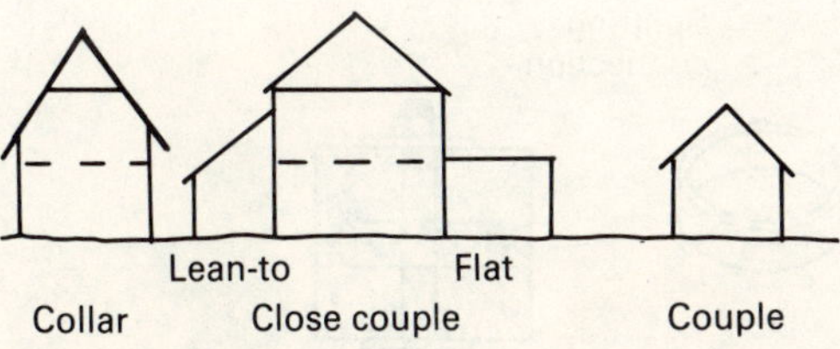

Fig. 9.13   Single roofs

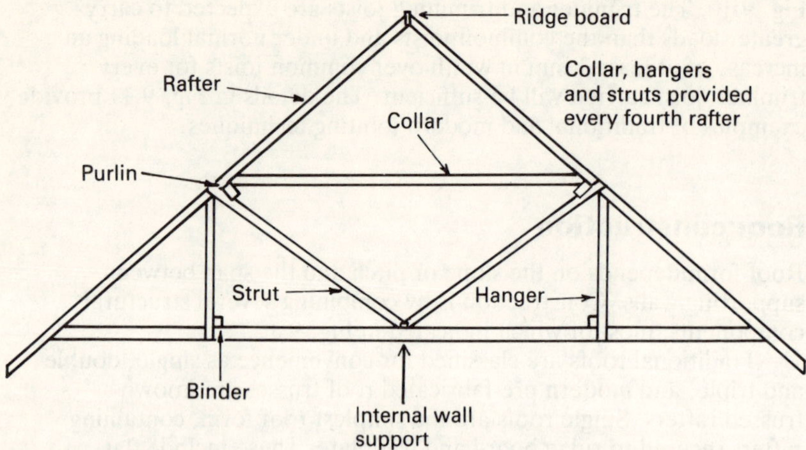

Fig. 9.14 Purlin or double roof construction

Additional members provide double roof construction using purlins with strut and possibly collar support at every fourth rafter. The exact arrangement of timber seldom varies and a possible double roof formation is shown in Fig. 9.14. Spans up to about 7.5 m are possible with vertical support below the purlin provided by hangers. Purlins run horizontally along the roof and their function is to resist deflection caused by the weight of roof cladding, snow and wind loading.

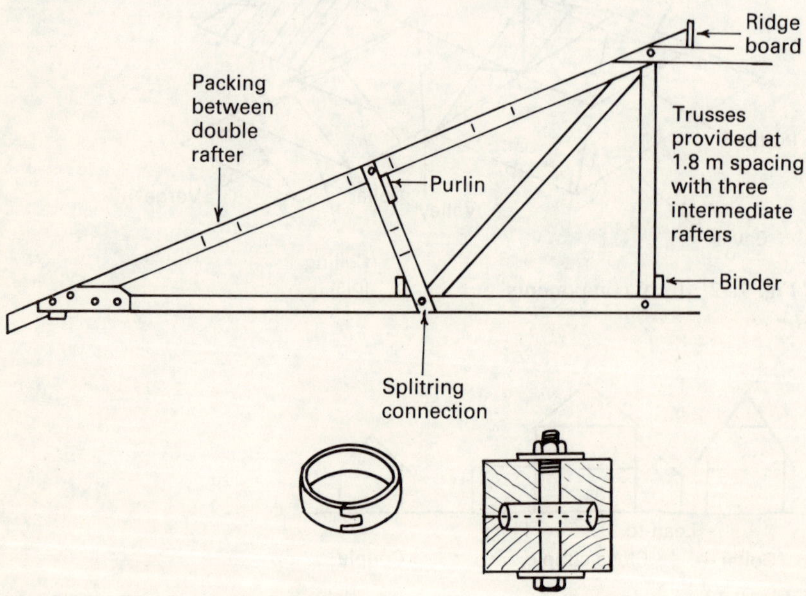

Fig. 9.15 TRADA type C domestic roof truss

Triple or framed roofs are composed of rafters and purlins with the addition of trussed support. The detail in Fig. 9.15 shows a truss which is used in a TRADA standard roof of this type. It is for domestic purposes (i.e. designed for tile coverings rather than sheeted covering) and it may be partially or completely assembled before delivery though it is usually built on site. Most of the joints use bolts with timber connectors which may be split-ring or tooth-plate connectors according to the loads to be met. Timber of structural grades is required. Maximum truss spacing is 1.8 m and there are usually three common rafters and ceiling joists between each truss so that their spacing is usually 450 mm. The TRADA roof can have reduced spacings and specially designed roofs can, of course, have any chosen spacing. This type of roof is well suited to spans between 5 and 11 m and the TRADA standard range lies within these limits. Designs for smaller and greater spans are, of course, possible.

Trussed rafters

Trussed rafters were introduced in Britain in 1962, following their successful use in North America since the early 1950s. They have many advantages over traditional roofs, primarily their clear span from wall to wall (removing the need for intermediate internal wall support), quality controlled factory manufacture, light-weight and simplicity of use by eliminating binders, ridgeboard and purlins.

Trussed rafters are produced from machined timber of uniform thickness, but jointed by means of toothed metal plates or steel plates with holes through which nails can penetrate the timber. See Fig. 9.16 for examples of these jointing plates.

Spacing of trussed rafters is a maximum of 600 mm and nailed direct to a timber wall plate at the head of the cavity wall. Vertically is provided by tiling battens secured with nails 32 mm longer than the batten thickness. Additional timbers running square to the trusses are secured to the ceiling joists and struts to prevent flexing on larger spans. Trussed rafters may be produced in a variety of geometric forms, the most popular being fink or fan arrangements shown in Fig. 9.17. Size of timber members varies with span and pitch, but sections will be smaller than those used for equivalent span traditional roofs, hence the importance of providing substantial bearers to support

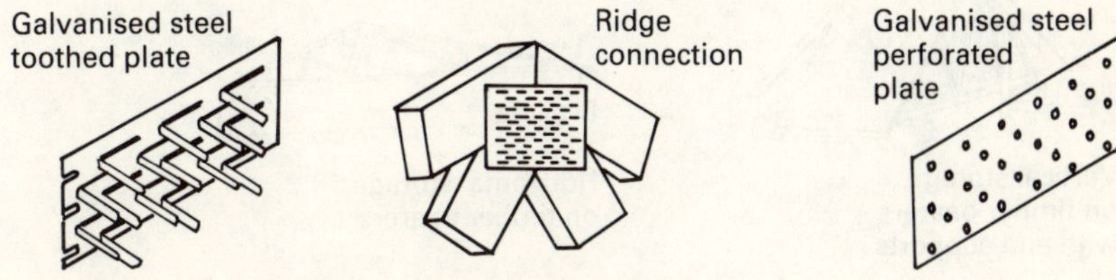

Fig. 9.16 Trussed rafter connector plates

134

water cisterns, as shown in Fig. 9.18.

Site handling of trussed rafters usually requires no more than two operatives and careful storage is essential to avoid damage to the timber and the plate connectors. Vertical storage is preferable, with the trusses supported where the wallplate would be situated and at sufficient height to allow rafter overhang to clear the ground. Alternatively, they may be laid flat on firm ground with bearers providing uniform and level support. Both methods are shown in Fig. 9.19.

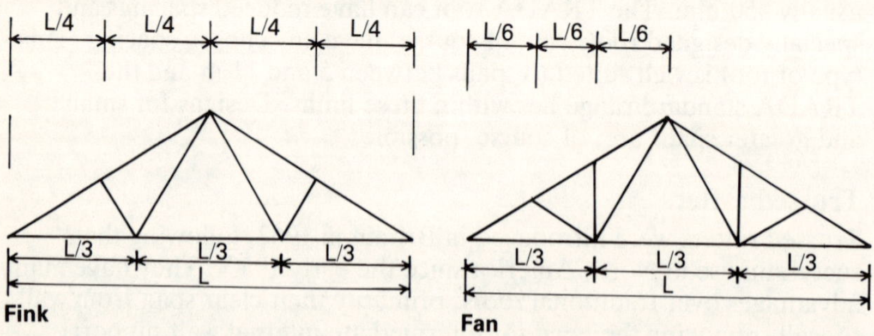

Fig. 9.17   Trussed rafters

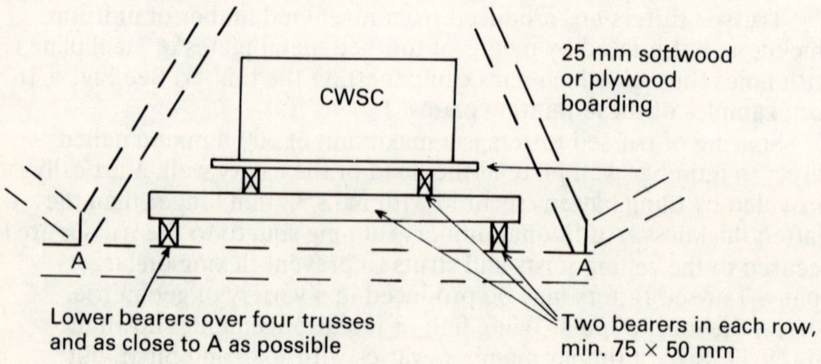

Fig. 9.18   Support to cold water storage cistern

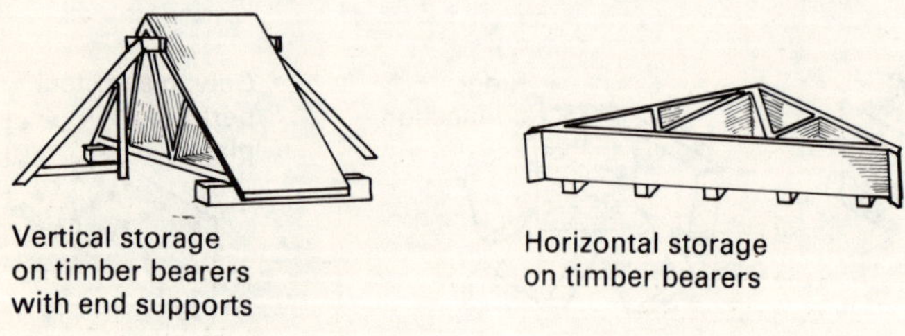

Fig. 9.19   Site storage of trussed rafters

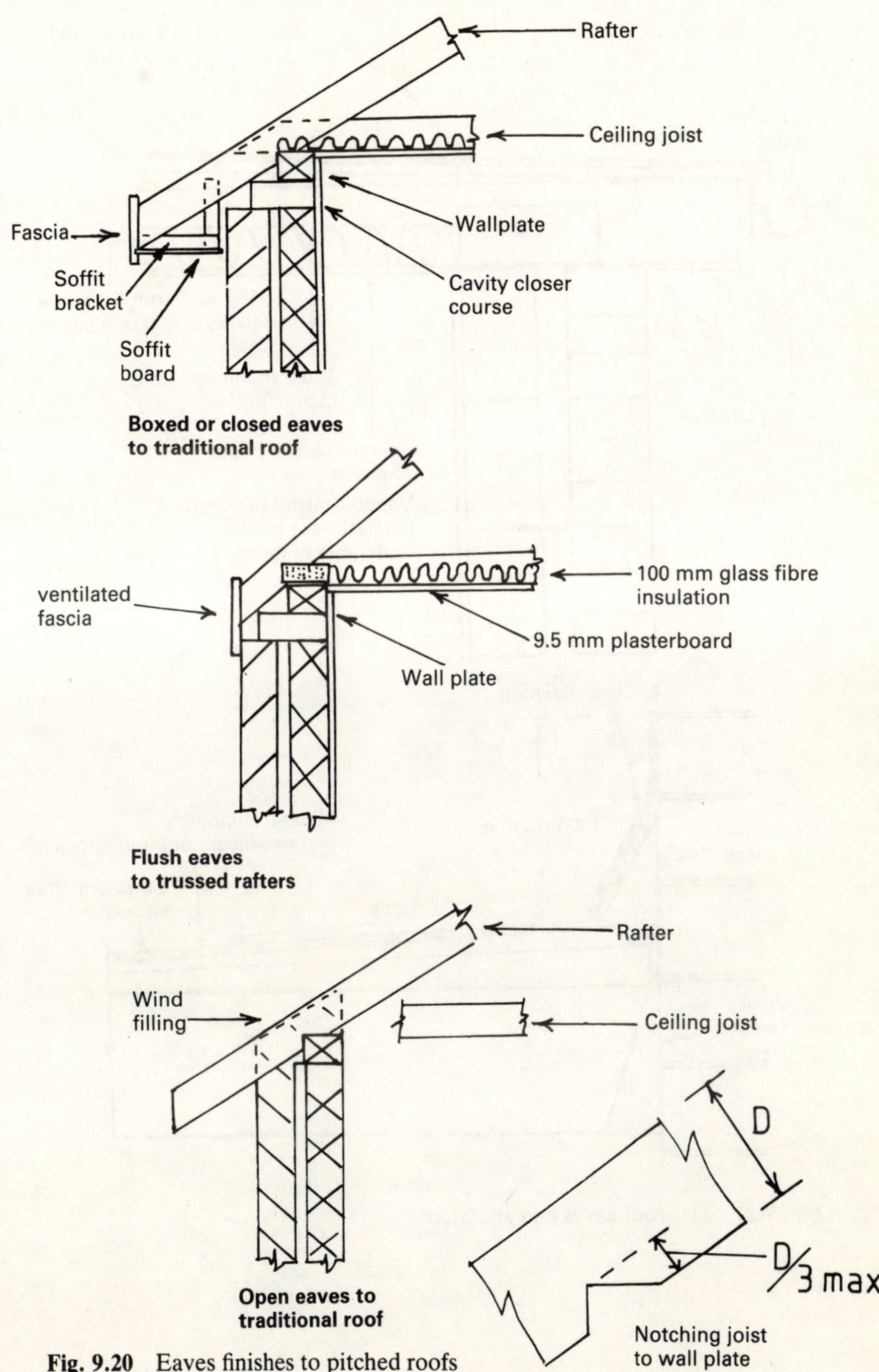

Fig. 9.20 Eaves finishes to pitched roofs

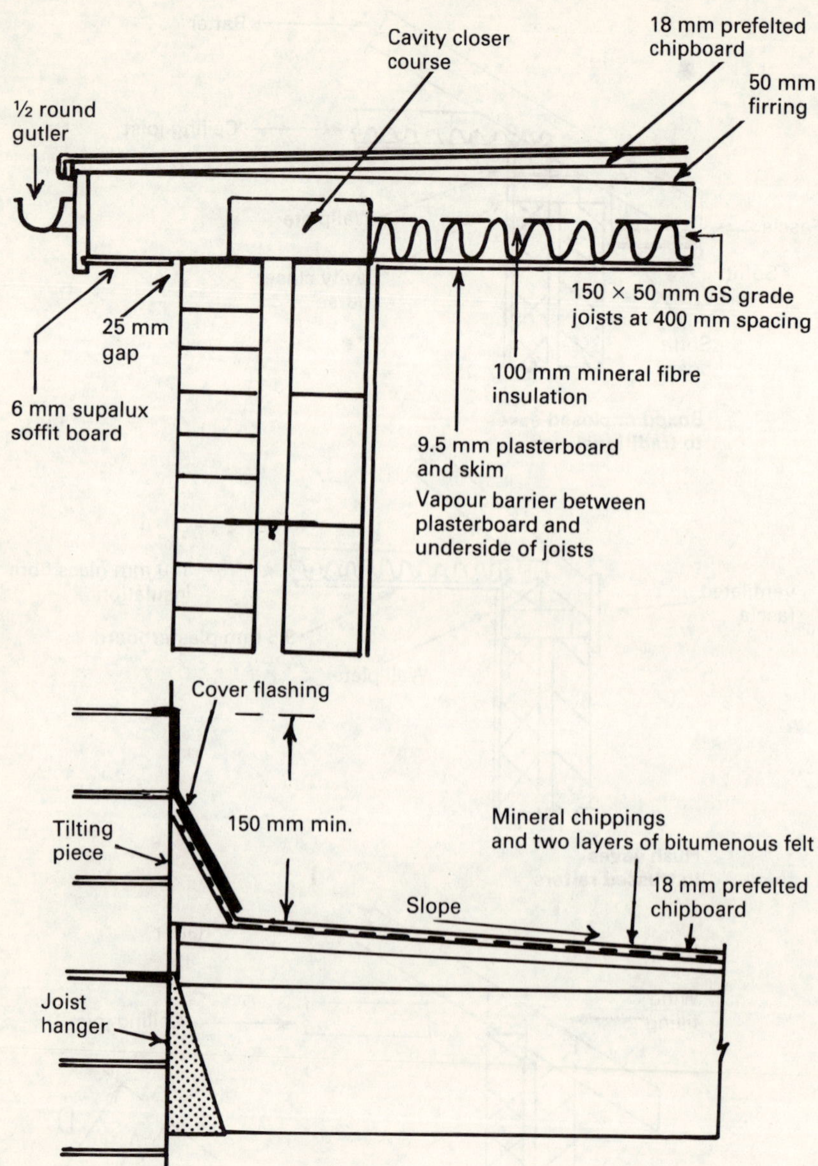

Fig. 9.21 Flat roof eaves and abutment

Eaves finishes

Construction at the eaves is normally boxed (or closed), open or flush. With each, the cavity wall is terminated with a block course sealing the cavity and a wallplate is provided to spread the roof load and provide a secure fixing for the roof. The details shown in Fig. 9.20 illustrate these finishes to a pitched roof. The treatment of a flat roof at eaves and abutment with an existing wall is shown in Fig. 9.21 whilst flat roof construction is shown in Fig. 9.22.

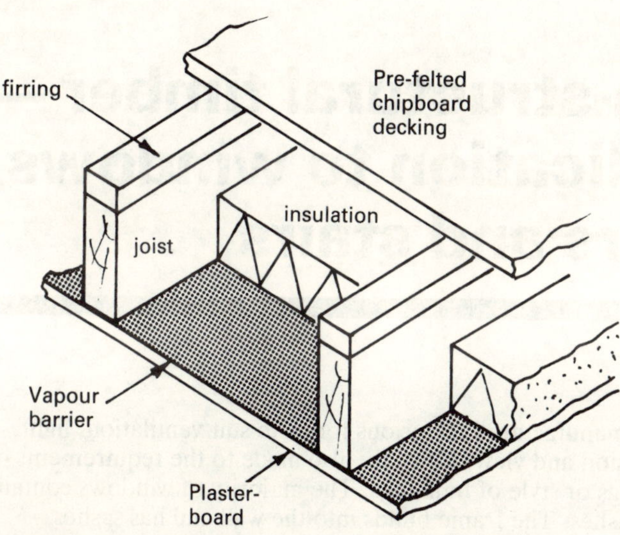

Fig. 9.22   Flat roof construction

Chapter 10

Non-structural timber — application to windows, doors and stairs

Windows

Windows are manufactured in various forms to suit ventilation, light, weather exclusion and view. They are also made to the requirements of special buildings or style of buildings. The majority of windows contain a frame and sashes. The frame builds into the wall and has sashes attached to contain the glass. The method of hanging or supporting the sash provides the different classification of windows. These are hinged or pivoted casement and sliding sash, but most joinery manufacturers sub-divide these into further categories, with variations such as bowed, georgian or jacobean.

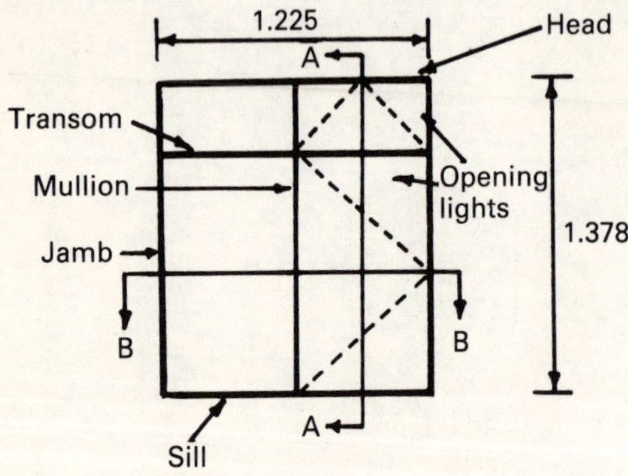

Fig. 10.1   Casement window. BS 644 ref. 246T

Figure 10.1 indicates the outline of a standard casement window indicating jamb, sill, head and intermediate members called transom and mullion. These are rebated to receive the sashes with anti-capillarity grooves provided where shown in Fig. 10.2. Frame components are usually produced in softwood, but hardwood is preferred for the more exposed sill. Sills have rebated and weathered upper surface with drip and mortar recesses to the underside. A 10 mm groove is also provided to the inside for location of window board.

Traditional vertical sliding sash windows have a long history, being very popular in pre-1930s houses and commercial buildings. Their construction is shown in Fig. 10.3, but the bulky nature of the boxing to the balance weights is undesirable and complicated: two sashes each slide in a pair of grooves and suspend by cords with lead, cast iron or concrete balance-weights. Modern sliding sash windows have similar appearance and operation to the traditional windows but undesirable boxing to sash weights is eliminated by provision of spiral spring balances contained within the sash. (see Fig. 10.4.)

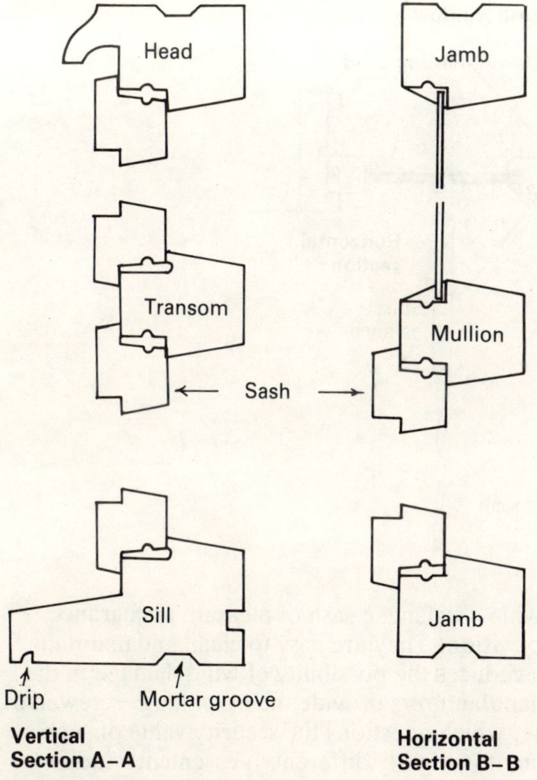

Fig. 10.2   British Standard casement window

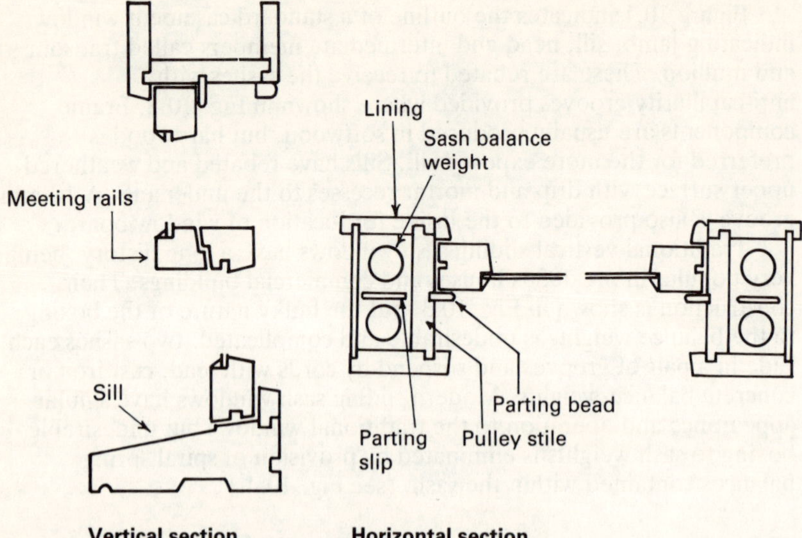

Fig. 10.3   Double hung sash window

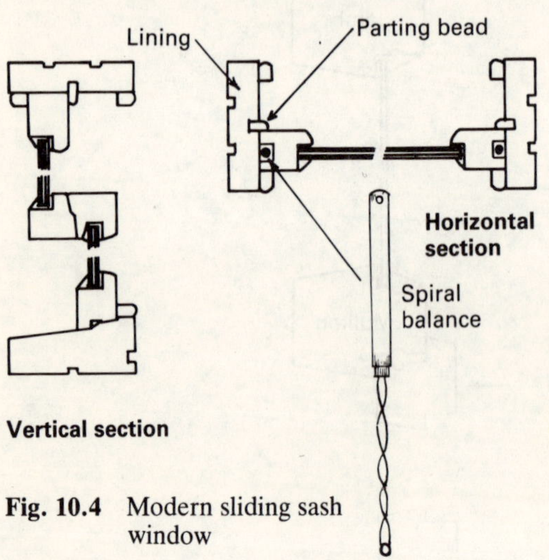

Fig. 10.4   Modern sliding sash window

Pivot windows provide a balanced sash of pleasant appearance offering a good range of vision. They are easy to clean and maintain, and the balanced effect reduces the possibility of wind damage in the open position. Some manufacturers provide the pivot hinge screwed to the outside of the frame, which questions the security value of such windows. Frame treatment is slightly different to casement window frames and a sectional detail is shown in Fig. 10.5.

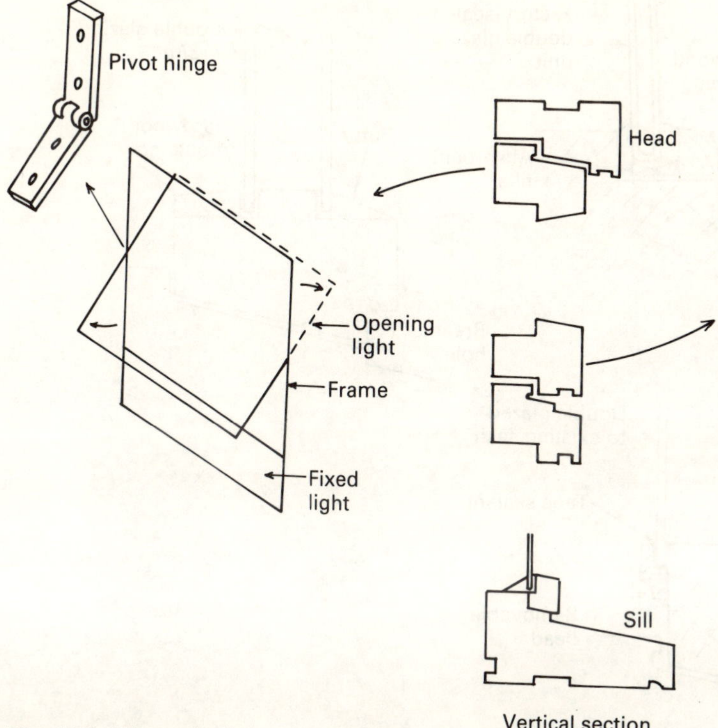

Fig. 10.5 Pivot window

Double glazing

Double glazing of sashes is an effective method of improving the
thermal insulation in heated buildings. Various methods are available,
including the simple 'do-it-yourself' attachments and hermetically
sealed units shown in Fig. 10.6. The optimum air space in between the
glass panes is between 12 and 20 mm; this reduces the heat loss to
approximately half that of single glazing. Increasing the space is of no
significance and reducing it below 12 mm worsens the insulation the
closer the panes become.

Doors

Doors are either side hung or have top and bottom sliding apparatus.
Their function is to provide privacy, security, weather exclusion,
suitable appearance and a reasonable degree of thermal insulation.
There are basically three categories of door: matchboard, panel and
flush.

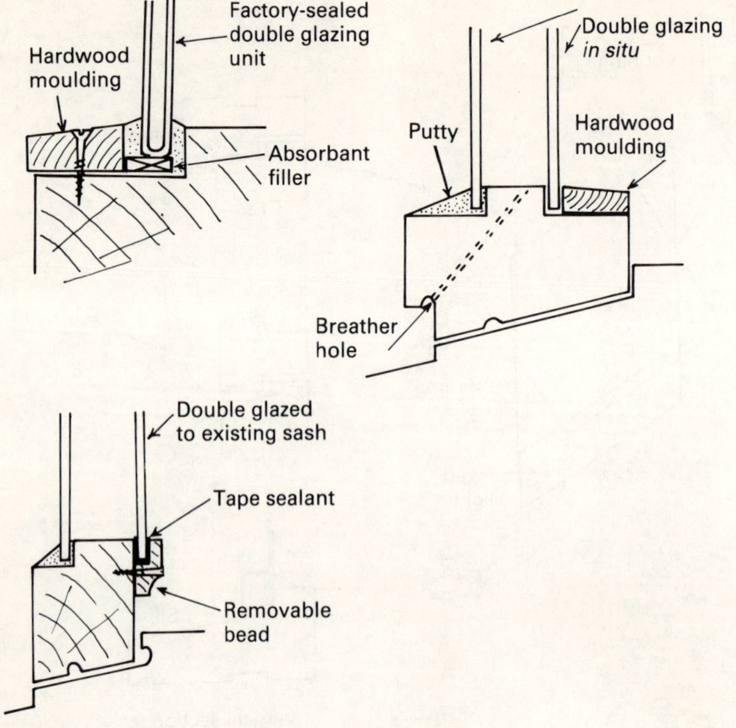

Fig. 10.6   Examples of double glazing

Matchboarded or battened doors

In their simplest form of battens and ledges this door is most suited to outbuildings, garages and huts. It is also found as internal and external doors to old cottages and farm houses. The tendency to sag and twist is reduced by bracing up from the hanging edge, and additional stiffening may be provided by an outer framework. The examples shown in Fig. 10.7 indicate BS 459 : Part 4 : 1965: *Dimensions of finished components*. Ironmongery consists of a pair of 'T' hinges, hasp and staple, bolt and door knobs.

Panelled doors

Panelled doors contain framing which is rebated to receive one or more panels of glass or plywood. The farming around the panel may be square, chamfered or moulded. If the panel is glazed, a shock-absorbant tape is provided between the glass and the frame fixing. The components of panelled door are illustrated in Fig. 10.8, and Fig. 10.9 shows BS 459: Part 1: 1954: Dimensions on a 2XG pattern door with position of butt hinges.

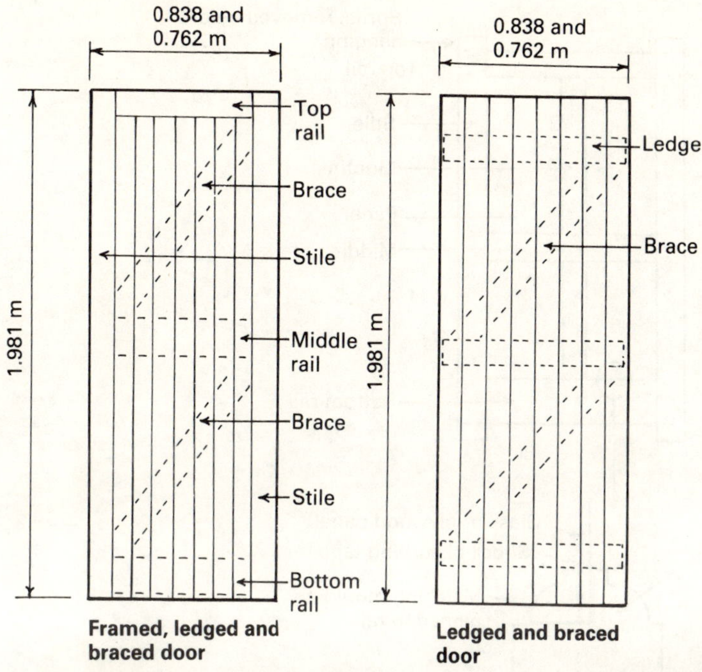

Fig. 10.7   Match boarded doors

Flush doors

Flush doors are very popular in most buildings as the finish and
construction may be varied to suit most functions. Finishing material
includes hardboard to BS 1142 : Part 1 : 1971, plywood to BS
1186 : Part 1 : 1971, exterior grade to be specified for external doors
and decorative hardwood veneers free from defects. Simulated
hardwood veneers produced by printing the hardwood grain and
pattern on hardboard are a cheaper alternative to real veneer, but
uniformity of finish is less desirable.

Construction depends on the purpose for which the door is
required. For sound insulation and fire resistance a solid core of
laminated timber is preferred, but for normal use in house
construction, the hollow core type is quite adequate. A semi-solid form
of construction, with internal filling of plasterboard or flaxboard, will
offer sufficient fire resistance in most situations. Examples of flush
door construction are shown in Fig.10.10. British Standard 459 : Part
2 : 1962 provides overall dimensions of 1.981 m in height for both
external and internal doors, interior door widths of 0.610, 0.686, 0.762
and 0.838 m, and exterior door widths of 0.762 and 0.838 m. Interior
door thickness equals 35 mm and exterior, 44 mm.

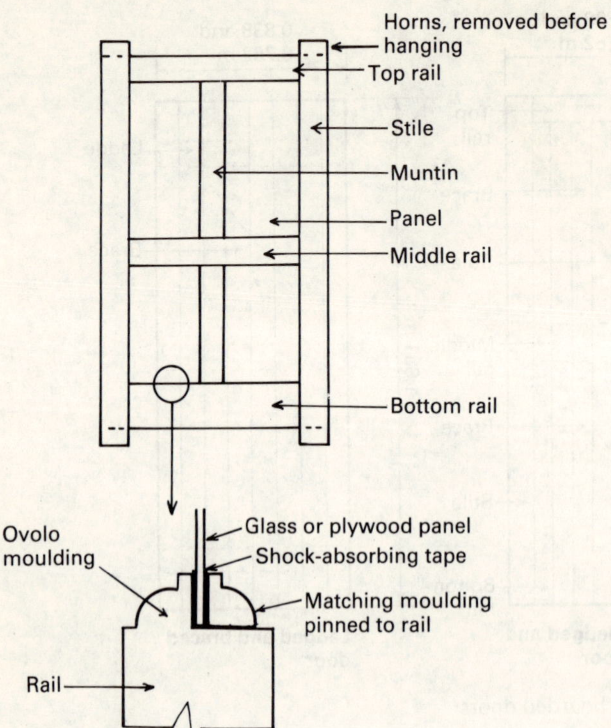

Fig. 10.8 Panelled door components

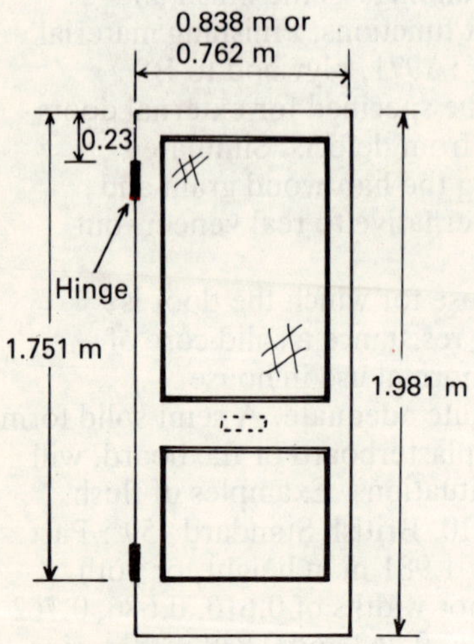

Fig. 10.9 Glazed panelled door, type 2XG

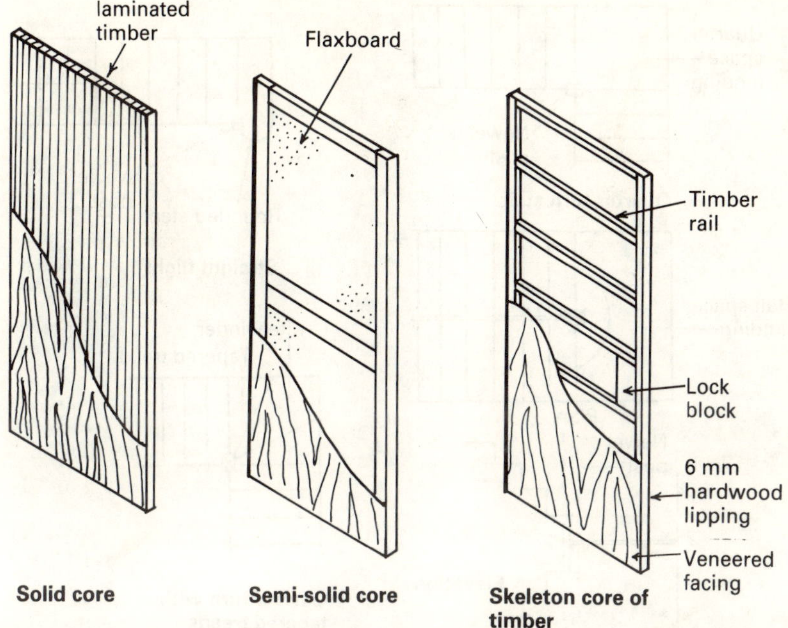

Fig. 10.10   Flush door construction

Stairs

Stairs to houses are manufactured from softwood with the possible use
of decorative hardwood as an ornamental feature. The form and
arrangement of stairway depends on the amount of space available,
and the effect the stairs are to have. In certain situations, particularly a
building designed purposely for aged persons, it is desirable to provide
landings. Some possible stairway plans are shown in Fig. 10.11.

Timber stairs have treads housed into strings, and these are
tenoned to newel posts supported by the floor structure. Glue blocks
are located at the junction of tread and riser, and on large and older
style stairways a carriage or bearer is provided to give support along
the centre of the stairway. Housed or closed strings have both upper
and lower edges parallel, the upper edge being sometimes capped to
receive balusters. The drawing in Fig. 10.12 provides a view of the
underside of the stair, illustrating the method locating treads and risers
to the string.

Site construction and assembly of a stair is only likely to occur
where the stairway is purpose-made to suit special circumstances. Most
joinery distributors provide ready-made standard stairs with treads and
risers or in open tread form as shown in Fig. 10.13. The Building
Regulations will only permit open tread stairs in private houses if a

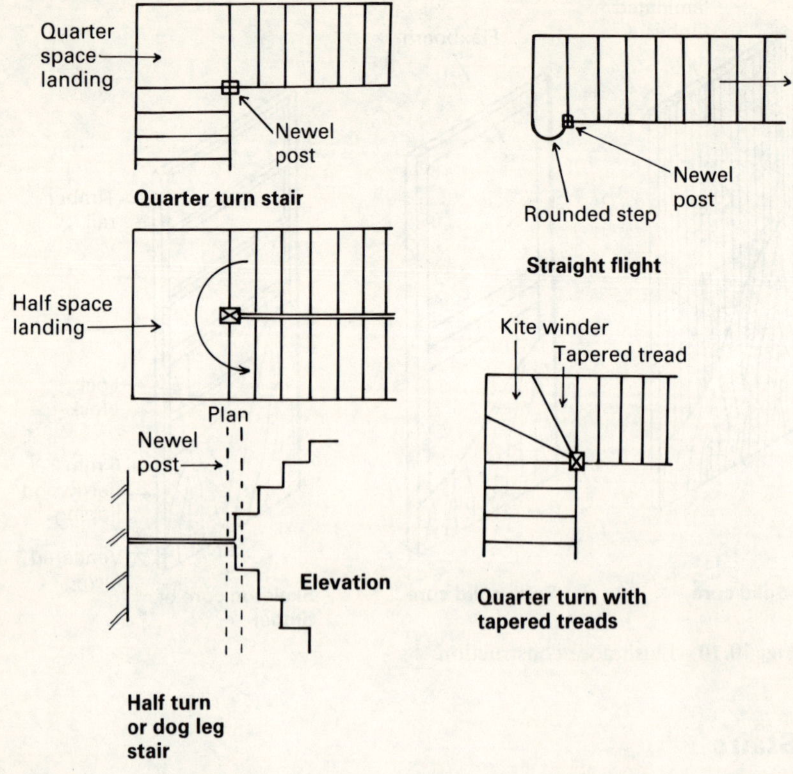

Fig. 10.11   Stair profiles

safety rail is provided beneath the tread to prevent a 100 mm diameter sphere passing through.

Part H of the Building Regulations applies very detailed control over the construction and use of stairways. All classes and types of building are affected, but for the purposes of this section consideration is limited to dwelling house construction. Regulations governing these stairs are illustrated in Fig. 10.14, which contains the following requirements.

Equal rise for every step or landing.
Equal going for every parallel tread.
The nosing of each open tread must overlap the tread below by at least 15 mm.
The maximum angle of the pitch line to the horizontal is 42°.
Going of tread must be no less than 220 mm.
Rise to be no more than 220 mm.
2 m minimum headroom measured vertically above the pitch line.
The sum of twice the rise plus the going must be equal to, or between, 550 and 700 mm.

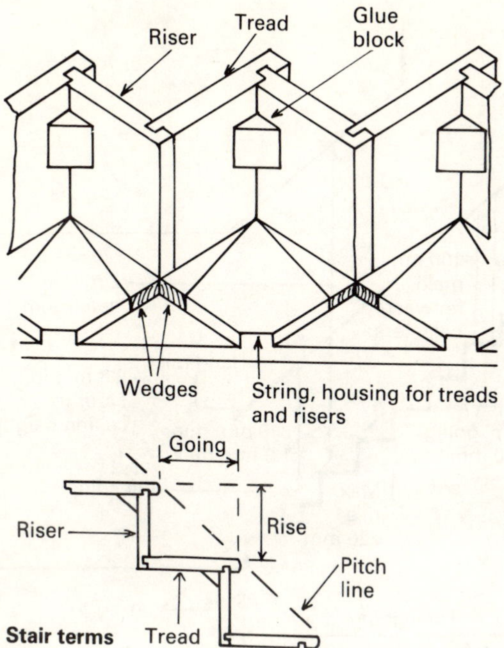

Fig. 10.12 Elevation of the underside of a stair showing housed treads and risers

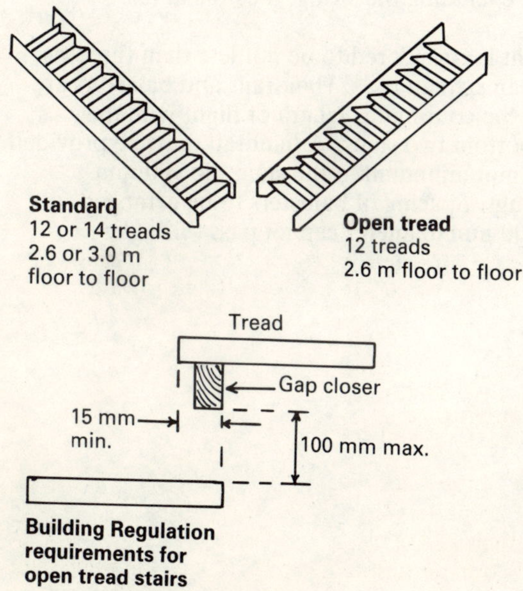

Fig. 10.13 Ready-made stairs

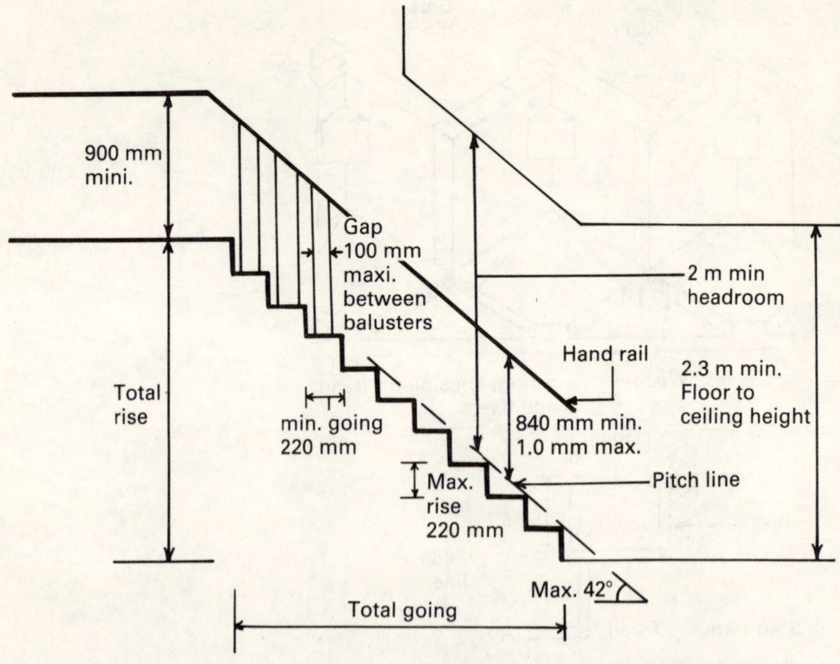

Fig. 10.14 Building Regulation requirements for domestic stairways

Unobstructed width, excluding the string, must be at least 800 mm.

Length of a stair flight is considered to be not less than three rises and not greater than sixteen rises. Handrails and balusters are required to guard the continuous length of flight with the exception of the bottom two steps. A handrail must be provided at a height of 840 mm minimum to the stair and 900 mm minimum to landings. Spacing of balusters to be arranged so that a sphere of 100 mm diameter cannot pass through.

Chapter 11

The temporary use of timber for hoardings and simple shuttering

Provision of fences and hoarding to a building site is enforced by the 1980 Highways Act, Sections 172 and 173. Work involving construction, demolition or repairs to a building requires a close boarded hoarding or a fence separating the building from the adjacent street. The form and construction must be to the local authority's satisfaction, and if the situation is suitable, the local authority are empowered to waive these requirements.

Where a hoarding is considered necessary, the local authority can insist that a temporary footway is provided outside the hoarding, having a handrail and suitable means of cover. It will also be necessary for the builder to maintain the hoarding in good and safe order, and if required provide white lighting beneath the cover, and red or hazard lighting around the perimeter. The builder will also be responsible for removal of the hoarding on completion of the contract and reinstatement of any damage to the footpath or highway. Failure to comply with this section of the Highways Act can incur a fine not exceeding £100 plus a further fine not exceeding £2 for every day the offence continues.

Section 173 requires the hoarding to be erected in a sound and stable state. Failure to recognise this can impose a maximum fine of £25 plus £1 for every day that the hoarding remains unsatisfactorily secured.

Selection of a suitable form of fencing or hoarding will depend on the location of the site and the nature of the work. In urban areas the local authority will probably insist on close-boarded hoarding, and will require a copy of the site plan with position of hoarding indicated by a coloured line. A specification and constructional detail of the proposed hoarding should also be submitted for approval. In rural areas, a simpler form of fencing or open hoarding is usually adequate to prevent trespass and possible accident to persons who may otherwise wander onto the site.

Chestnut pale fencing

Chestnut pale fencing is a light, re-usable fencing system available in rolls which simplifies erection and dismantling. This type of fence is shown in Fig. 11.1, with pales spaced at between 50 and 100 mm. Pales are connected by two or three lines of galvanised steel wire and different fence heights are available between 0.9 and 1.8 m. Vertical support is secured by tying to round chestnut posts, provided at 2 to 3 m spacing. This type of fencing is most suited to rural situations and uncongested areas. Its effect is sufficient to provide an obvious barrier against trespass, but it is of limited use where security against theft of materials is of major significance.

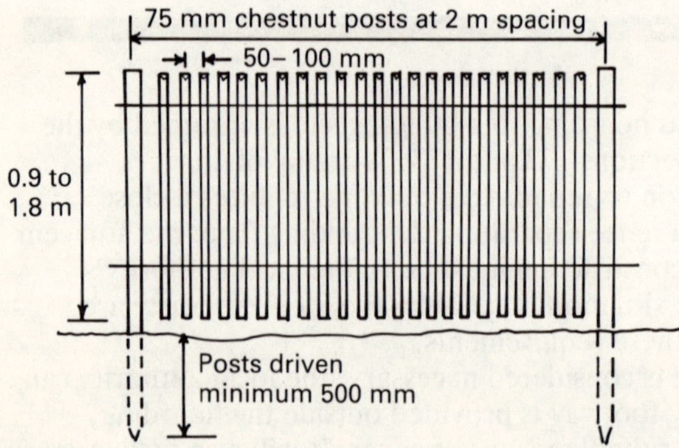

Fig. 11.1   Chestnut pale fencing

Close boarded hoarding

Close boarded hoarding is much more satisfactory for protecting the site against intruders, and as a barrier for the safety of the general public. The hoarding materials most suited are exterior grade plywood or corrugated steel sheeting attached to timber horizontal rails and vertical posts. Posts are driven into the ground for a firm fixing or provided with a shallow excavation and located with a bed of concrete. The detail in Fig. 11.2 indicates a typical timber hoarding with bracing to provide rigidity and resistance to wind loading.

An alternative support to hoarding is provided by scaffolding to the building. Here, timber rails bolt direct to standards as shown in Fig. 11.3, and a fan guard or fan hoarding inclines outwards over the

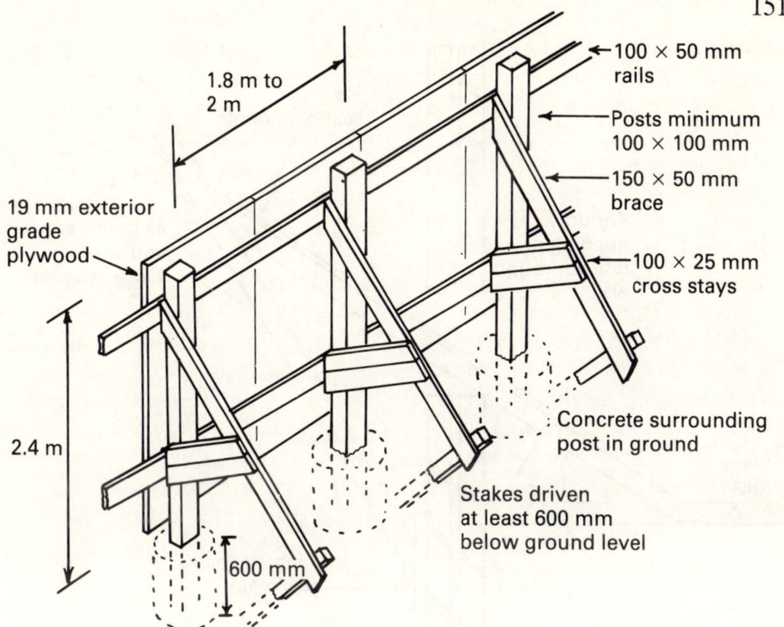

Fig. 11.2 Free standing close boarded hoarding

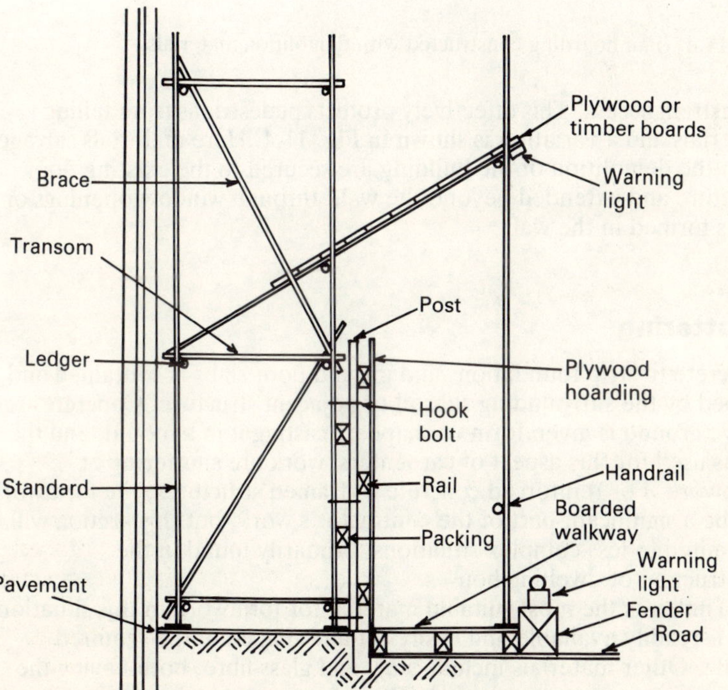

Fig. 11.3 Hoarding with support from scaffolding

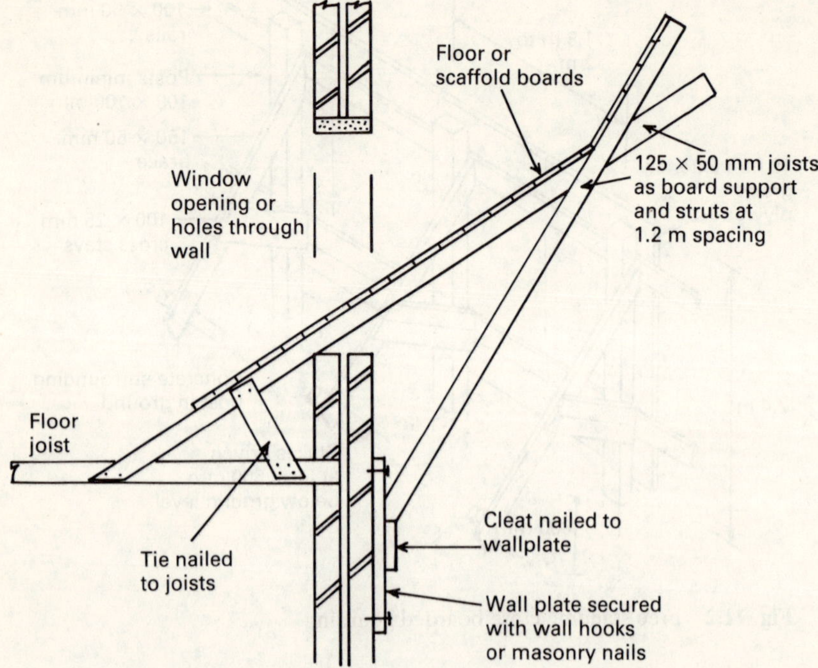

Fig. 11.4   Fan hoarding constructed with demolition materials

pedestrian access. This effectively protects pedestrians from falling materials and a variation is shown in Fig. 11.4. Here materials salvaged from the demolition of the building are secured to the existing floor structure and extended beyond the wall, through window openings or holes formed in the wall.

Shuttering

Concrete to strip foundations and ground floor slabs is contained and shaped by the surrounding subsoil or adjacent structure. Concrete above ground is given form or shape by casting it in a mould, and the terms used for this aspect of carpenters' work are shuttering or formwork. For reinforced concrete or framed structures, the formwork will be a significant part of the contractor's work, but this section will be limited to less complex situations, primarily found in the construction of dwelling houses.

Timber is the most suitable material for formwork in this situation, as it is readily available and easily cut and shaped to the required profile. Other materials include steel and glass fibre, both having the advantage of numerous re-uses. These forms are more suited to repetitive construction on a uniform matrix of the type associated with car parks and multi-storey buildings.

Sawn timber will leave a rough finish to the concrete surface. This is an advantage where plaster is expected to bond; however, where the concrete unit is exposed a smooth finish is often more acceptable. This may be achieved by using planed timber or lining the mould with hardboard or plywood. Rough timber may also encourage the concrete to adhere to its surface, disturbing the surface finish when the formwork is struck. To avoid this a patented mould oil painted over the inside of the form will prevent concrete adhesion and moisture from the concrete penetrating the timber.

Formwork must be of sufficient strength and rigidity to withstand the dead load of the concrete. Additionally it must have sufficient strength to permit pouring, vibrating and tamping of the concrete without movement. Uniform sections such as lintels, fence posts, kerb stones or sills are usually pre-cast and lifted into place. This eases the construction process, eliminating the need for inconvenient and obstructive shuttering about the structure. Thoughtful planning will be necessary to allow the concrete several days to harden, before

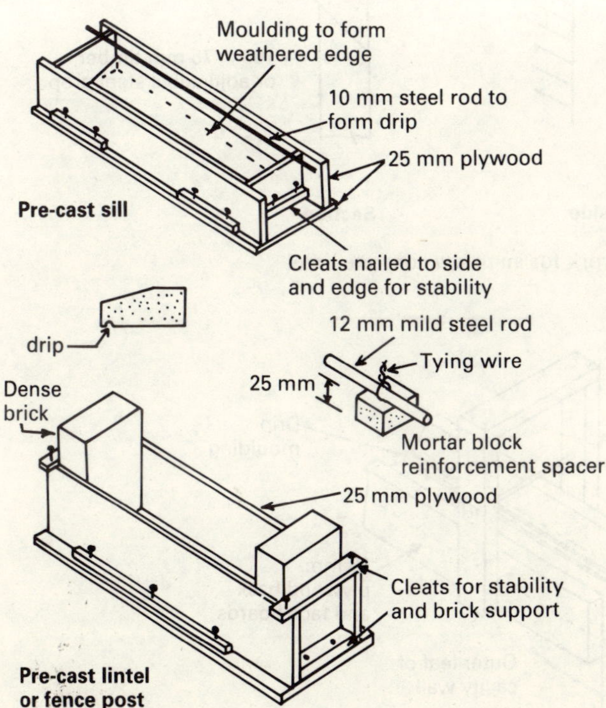

Fig. 11.5 Formwork for simple uniform sections

removing formwork and placing the unit. Examples of simple formwork to uniform sections are shown in Fig. 11.5.

Many lintels are expected to span large openings. To avoid the danger and difficulty of manouevering a large beam into place, formwork is constructed at the head of the opening. The example shown in Fig. 11.6 represents a suitable formwork system for this situation. Also, it is often more convenient to cast sills *in situ* by clamping the formwork to the wall, as shown in Fig. 11.7.

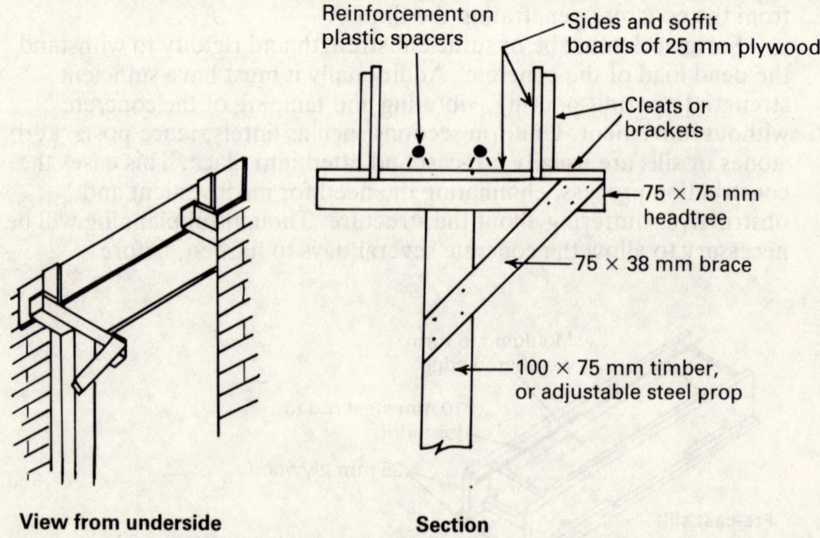

Fig. 11.6 Formwork for an *in situ* concrete lintel

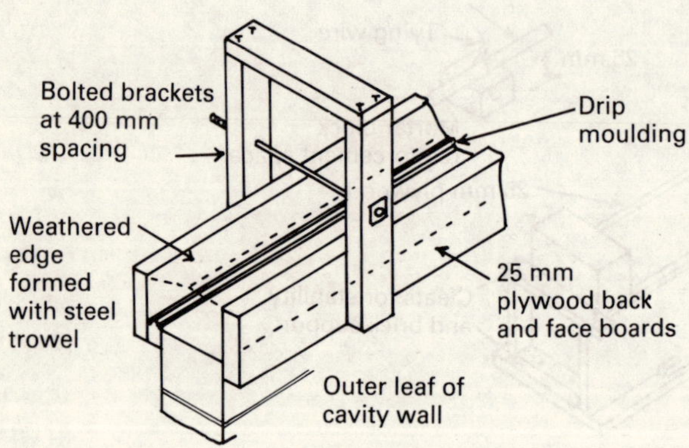

Fig. 11.7 Formwork to cast *in situ* sill

Cast *in situ* concrete ground floor slabs are ideally contained by the perimeter brickwork acting as a permanent formwork. Where the external wall is not available, as with timber-framed housing, shuttering of the type shown in Fig. 11.8 will be suitable. Here 75 × 75 mm timber soldiers or vertical posts and 25 mm plywood are restrained with a ground plate, struts and spreading timber.

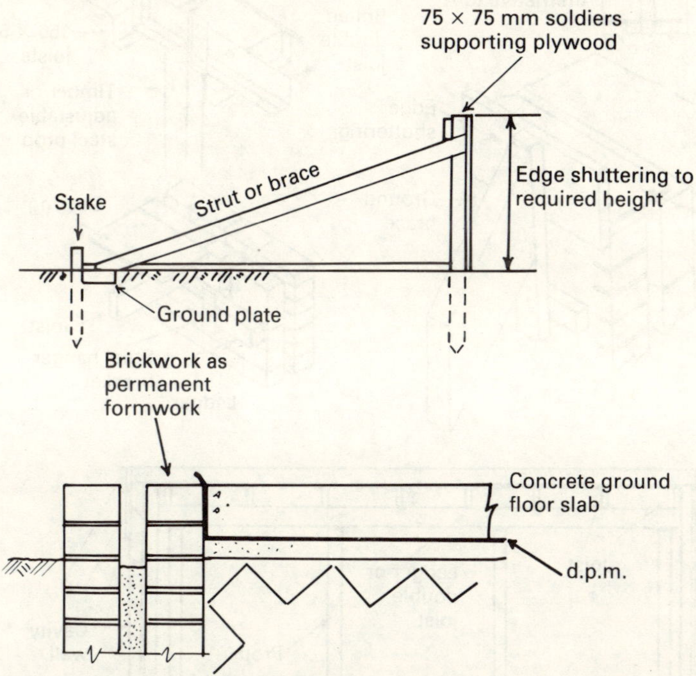

Fig. 11.8   Formwork to ground floor slabs

Concrete upper floors and flat roofs to dwelling houses are an unusual form of construction, but are possible as sound- and fire-resisting floors between flats, or between a shop and a flat. The formwork contains 19 mm or preferably 25 mm plywood decking to receive the concrete, supported by 150 × 50 mm joists at between 400 and 600 mm spacing, depending on the deck thickness. Shuttering grade plywood is normally specified for this work, having one side only prepared. Joists are supported by ledgers, usually two 150 × 50 mm joists bolted together and spaced at a maximum of 1.5 m. Vertical props of timber of adjustable steel props provide ground support to the ledgers at between 1.5 and 2 m spacing, depending on the mass of concrete to be supported. A typical arrangement is shown in Fig. 11.9, with the method of providing ledger and joist support to the decking. Edge detail is also included.

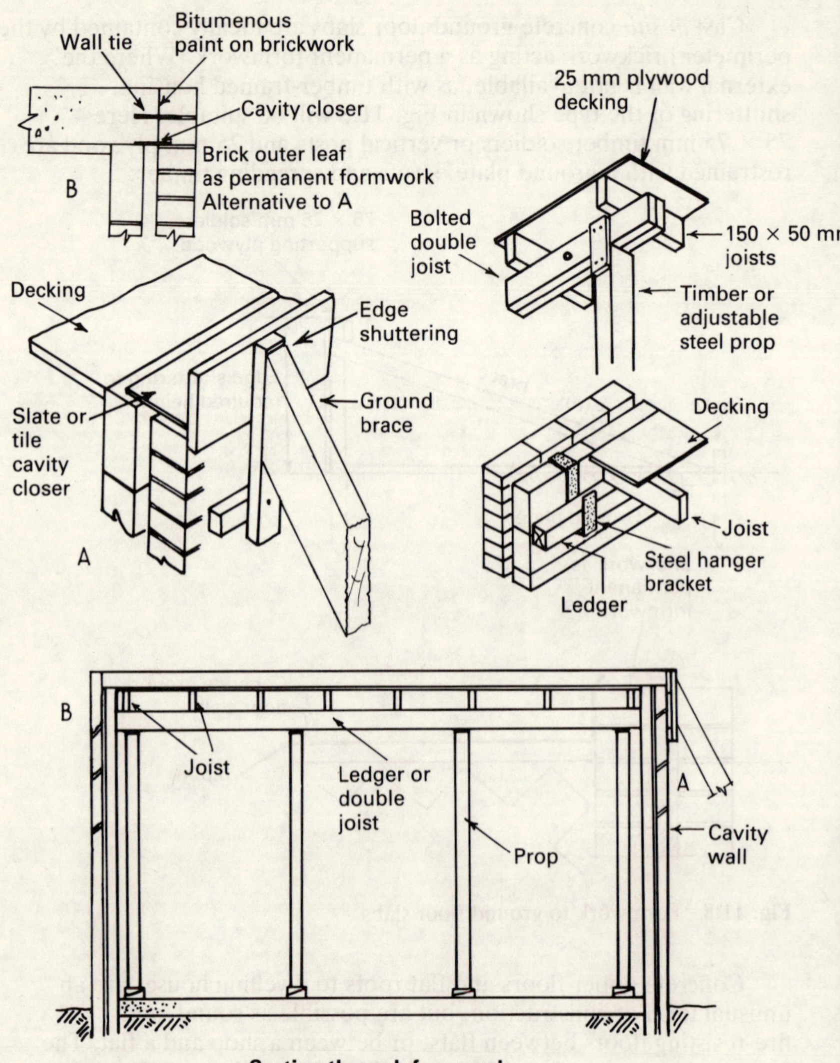

Fig. 11.9 Upper floor formwork

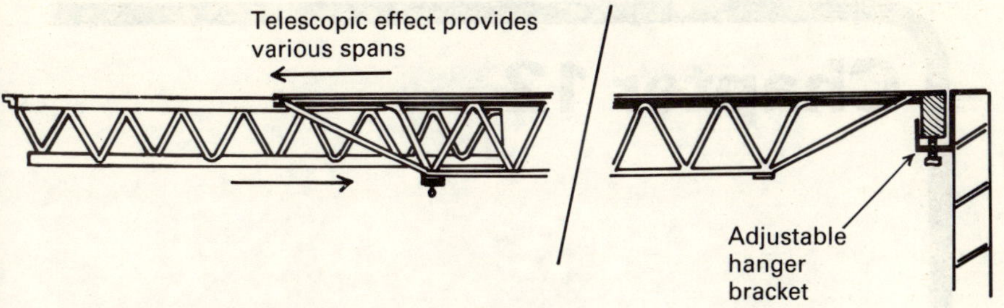

Fig. 11.10   Adjustable steel formwork supports

An alternative which will economise in the use of timber and formwork construction time, particularly with larger spans, is lightweight adjustable floor centres. These substitute for ledgers and eliminate the use of props, providing a clear working area below the formwork. They are normally located at 600 mm spacing, but spacing up to 1.2 m is possible where loads are not excessive. Adjustable centres have considerable re-use value, and contractors requiring them at intermittent periods may prefer to hire rather than commit capital to their purchase. Application is shown in Fig. 11.10.

Chapter 12

Reinforced concrete construction

Concrete

The basic components and practical application of concrete are explained in Chapter 6. This chapter is primarily concerned with the use of steel reinforcement in concrete structural units.

Concrete may be produced by prescribed or designed mix. Prescribed mixes are a combination of known quantities of water, cement and aggregates to produce concrete of a predetermined minimum strength. The actual strength will vary between batches and is often far greater than required, signifying the waste of some components. Variation is determined by method of mixing and experience of the person producing the concrete. Designed mixes are produced by experience and knowledge of the mix components and mixing machinery. They are more economic in materials, particularly cement, and different batches will produce constant strength results, hence the reliability of ready-mixed supplies.

At 28 days, concrete has achieved 75 per cent of its maximum strength, the remaining 25 per cent obtained over a period of years. The compressive strength of concrete at 28 days is known as the characteristic strength and BS, CP 110 : Part 1 : 1972, provides nine grades corresponding with characteristic strength. These are 7, 10, 15, 20, 25, 30, 40, 50 and 60 (N/mm^2). Concrete for normal reinforced structural purposes will be grade 20, 25 or 30. Lower is too weak for use with reinforcement and higher grades are more suitable for post-tensioned and pre-tensioned reinforcement.

Steel bar reinforcement

The types of steel normally used as reinforcement to concrete are mild steel or high-yield steel. Both steels contain about 99 per cent iron, the remaining 1 per cent is composed of manganese, carbon, sulphur and phosphorus. The ratio of minor components determines the quality of steel, mild steel having 0.25 per cent carbon and high-yield steel 0.4 per cent. Alternatively, high-yield steel may be produced by cold working or deforming ordinary mild steel until it is strain hardened. Various examples of steel reinforcement are contained in Fig. 12.1; some are ribbed to increase bond or adhesion to the surrounding concrete.

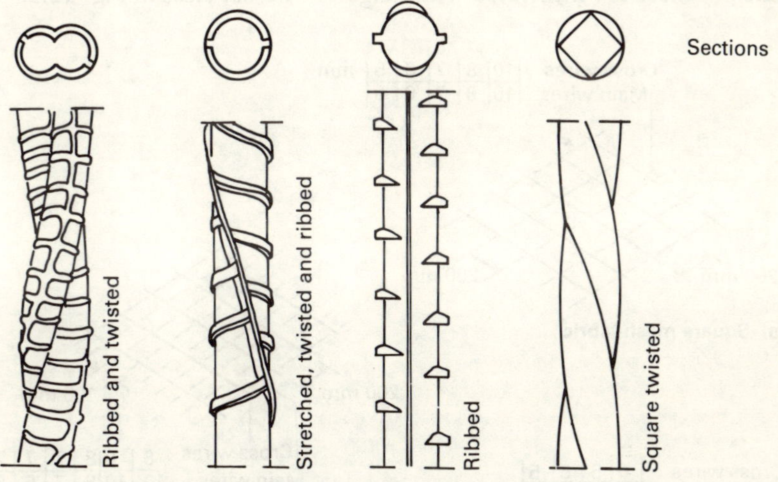

Fig. 12.1 Deformed steel reinforcement

British Standard 4449 : 1978 specifies hot-rolled steel bars as grade 250 or 460/425. These figures refer to the specified characteristic tensile strength in N/mm^2. Grade 250 is the acceptable standard of plain round mild steel bars containing a maximum of 0.25 per cent carbon, 0.06 per cent sulphur and 0.06 per cent phosphorus. Grade 460/425 specifies high-yield steel of maximum 0.4 per cent carbon, 0.05 per cent sulphur and 0.05 per cent phosphorus. 460 N/mm^2 refers to the specified characteristic strength of bars up to 16 mm, and 425 N/mm^2 to bars over 16 mm. British Standard preferred reinforcement sizes are 8, 10, 12, 16, 20, 25, 32 and 40 mm.

Steel mesh or fabric reinforcement

In addition to bar reinforcement, steel bars or wires are made available in welded mesh form for convenient reinforcement to concrete floor slabs and roads. These are also suitable for wrapping structural steel sections to provide binding for concrete fire protection, and are

160

available in long or square mesh form to BS 4483 : 1969. The BS categories include:

BS reference	Type	Use
A	Square mesh fabric	Floor slabs
B	Structural mesh fabric	Floor slabs
C	Long mesh fabric	Roads and pavements
D	Wrapping fabric	Sprayed concrete work and wrapping to steel sections, before concreting.

Each of these are illustrated with range of wire/bar sizes in Fig. 12.2.

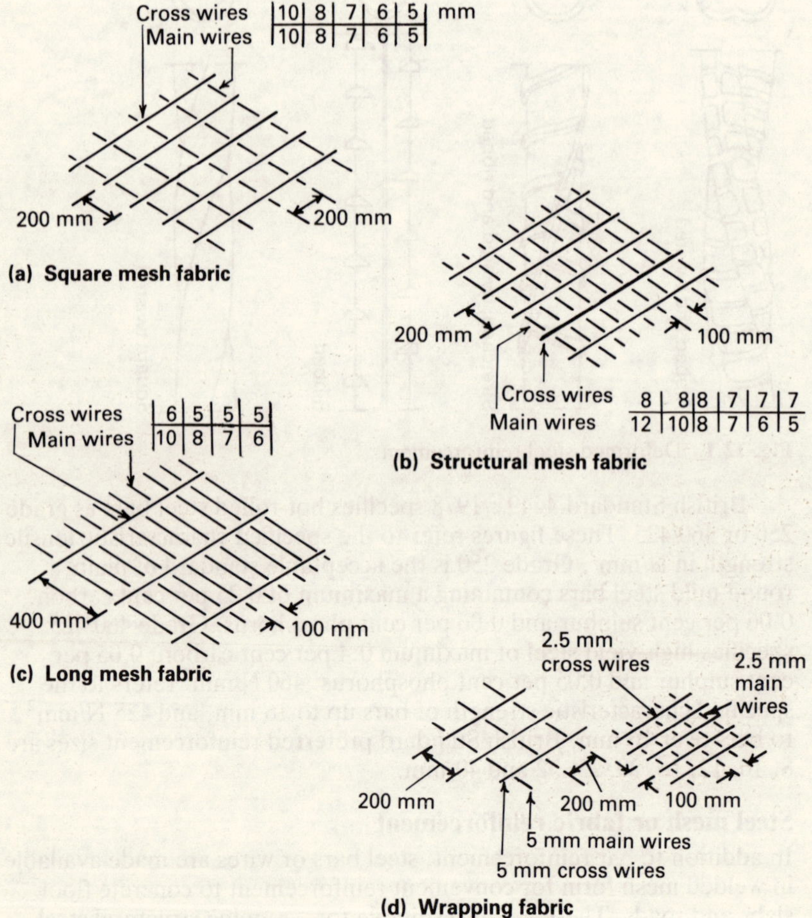

Fig. 12.2 British Standard Mesh

Fabrication of steel reinforcement

Preparation of steel reinforcement by cutting and bending on site is unusual, except where the reinforced concrete work is minimal. For convenience, particularly on large reinforced concrete framed projects, it is preferable to receive all steel preformed to the structural engineer's requirements. To simplify and to co-ordinate prefabrication, BS 4466 : 1981 provides a clear, unambiguous system for scheduling steel reinforcement. Numerous possible shapes are illustrated in the British Standard and defined by a code number. A portion of the British Standard has been extracted to illustrate this in Fig. 12.3.

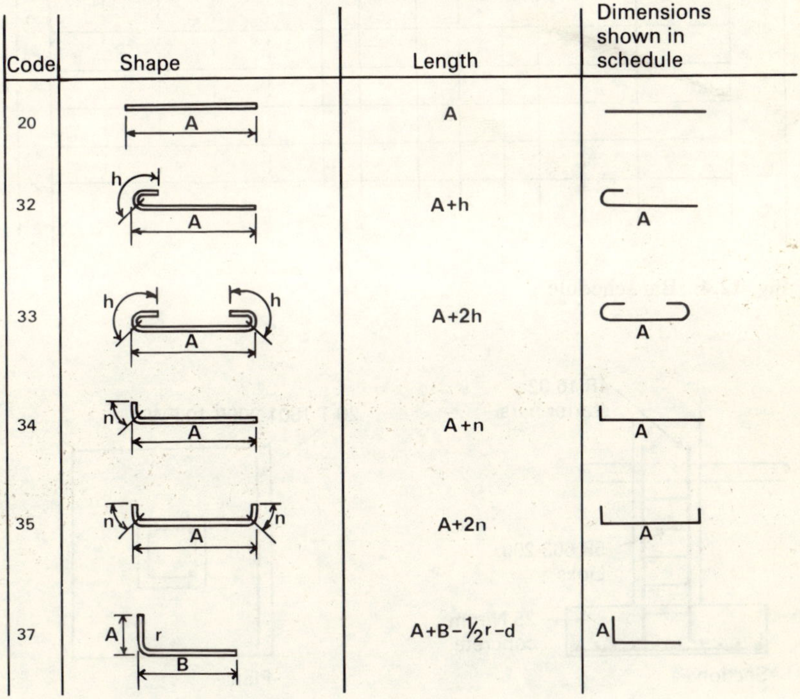

Code	Shape	Length	Dimensions shown in schedule
20	A	A	
32	A, h	$A+h$	A
33	A, h, h	$A+2h$	A
34	A, n	$A+n$	A
35	A, n, n	$A+2n$	A
37	A, B, r	$A+B-\tfrac{1}{2}r-d$	A

Fig. 12.3   Bar bending extract from BS 4466

Variable factors such as bar length and type of steel are included on a bar schedule as shown in Fig. 12.4. This simplifies and rationalises the cutting, bending and ordering of materials. It is also very useful on site when used in conjunction with the structural engineer's drawing. The use of this schedule is shown with the simple reinforced concrete pad foundation to a square column shown in Fig. 12.5. Figure 12.6 shows a possible suspended reinforced concrete flat slab with abbreviated notation for reinforcement. Taking 3R10–2–300T as an example,

Site ref:

Bar schedule ref: □□□ □□ □

Date prepared

Prepared by

Member	Bar mark	Type and size	No of mbrs	No. of bars in each	Total no.	Length of each bar mm	Shape code	A mm	B mm	C mm	D mm	E/r mm
Pad foundation	01	T16	4	20	80	1350	35	1100				
	02	R16	4	4	16	1200	37	100				
	03	R6	4	5	20	1300	60	300	300			

Fig. 12.4   Bar schedule

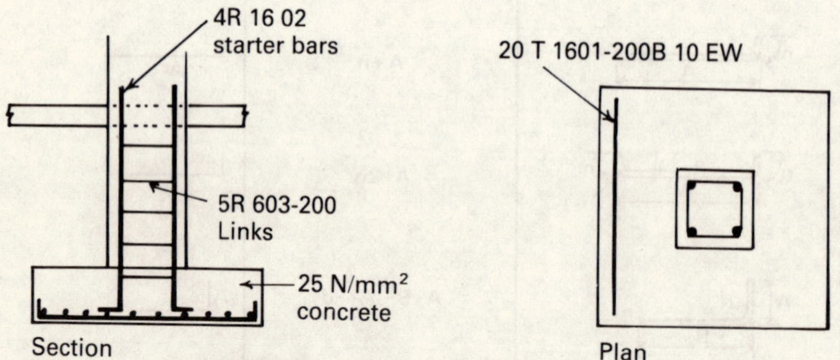

Fig. 12.5   Reinforced concrete pad foundation 4 No. required

    3 = quantity or number of bars.

    R = mild steel bars to grade 250 (T = h−y steel)

  10 = 10 mm diameter

    2 = bar mark or reference number

300 = spacing in millimetres

    T = placed in the Top.

Bars of the same shape code, type and size are bound together and labelled with the bar schedule reference and bar mark number.

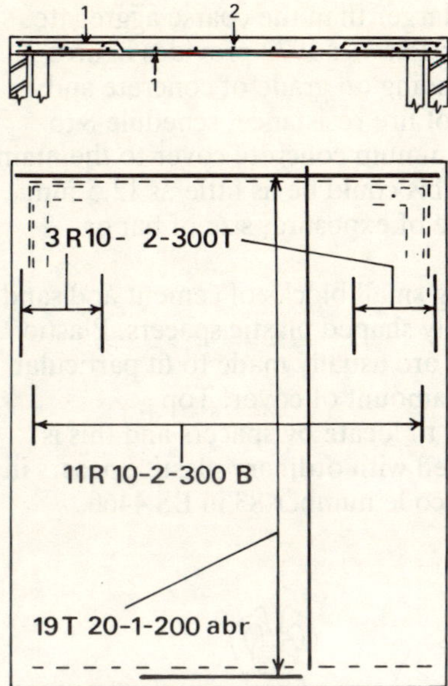

Fig. 12.6 Suspended reinforced concrete slab

End hooks

End bends and hooks can often be eliminated where ribbed
reinforcement is employed. Where they are used, the method of
forming is shown in Fig. 12.7. They have an internal radius of bend
conforming to twice the bar diameter for grade 250 steel and three
times the bar diameter for grade 460/425 steel. Dimensions n and h
vary with size of bar, but amount of straight length after the bend
should at least equal four times the bar diameter.

Concrete cover to reinforcement

Cover to reinforcement is determined by degree of fire resistance
required and extent of exposure to the weather. The cover should be at

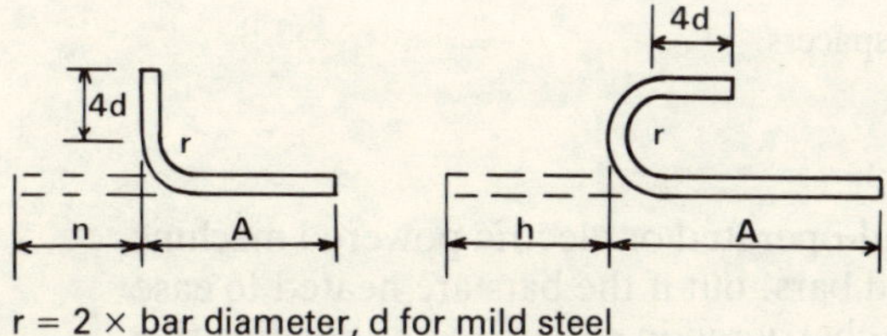

r = 2 × bar diameter, d for mild steel
   3 × bar diameter, d for high yield steel

Fig. 12.7 Reinforcement end hooks

least equal to the size of the bar and larger than the coarse aggregate. Table 19 to BS Code of Practice 110 : Part 1 : 1972 provides figures ranging between 15 and 60 mm depending on grade of concrete and condition of exposure. For purposes of fire resistance, schedule 8 to the Building Regulations provides minimum concrete cover to the main reinforcement in beams and lintels. This could be as little as 12.5 mm for half an hour resistance, but degree of exposure, size of bar or aggregate are likely to require more.

Correct cover is obtained by using small blocks of cement and sand mortar in the ratio of 1 : 2, or specially shaped plastic spacers. Plastic spacers are now commonly used, and are usually made to fit particular bar sizes to provide a predetermined amount of cover. Top reinforcement in a slab is too difficult to locate by spacers and this is supported on 'chairs'. This is illustrated with ordinary plastic spacers in Fig. 12.8. The chair shown is bent to code number 83 in BS 4466.

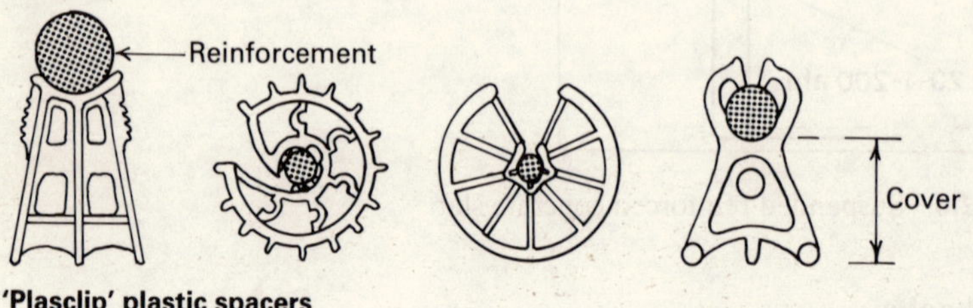

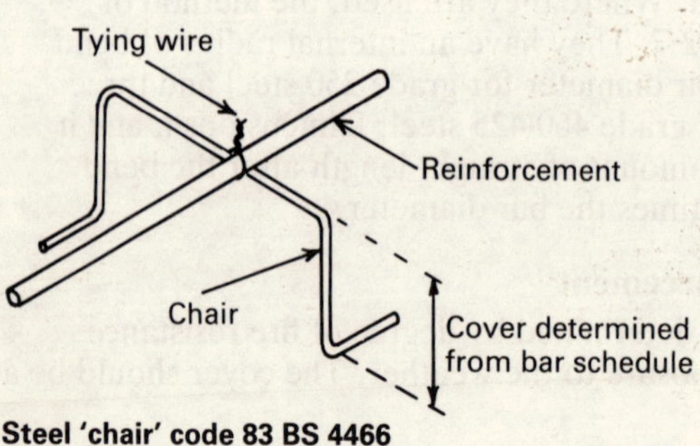

Fig. 12.8 Reinforcement spacers

Bending and cutting

Bending is either by hand-operated or electric-powered machine. Bends are formed in cold bars, but if the bars are heated to ease bending, they should be cherry red in colour. After bending they should be allowed to cool slowly, a quenching will induce hardening. The drawing in Fig. 12.9 illustrates a typical bench bender, with radius

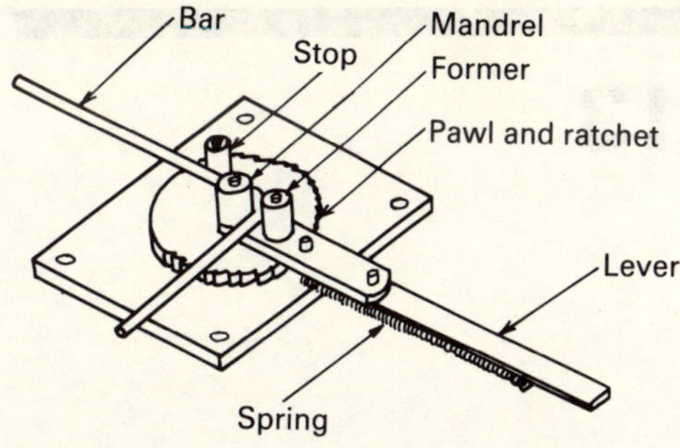

Fig. 12.9   Bench-mounted bar bender

of bend formed around the centre mandrel. Cutting is normally by bench-mounted guillotine, but hacksaw or bolt croppers are adequate for smaller size bars. Cutting by oxy-acetylene equipment is also possible and this can be useful for minor adjustments and welded connections to smaller size bars.

Reinforced concrete

Steel has excellent compressive and tensile strength properties, but has the disadvantage of high cost. Furthermore its performance in fire is poor, therefore cladding of suitable protective materials is essential in most situations. When exposed it will rust, and again protection is necessary. Concrete, however, is relatively cheap, behaves well in fire and is very durable. It is weaker than steel when subject to compressive force, but more than adequate for normal structural use. In tension it is very much weaker and will fail at about 10 per cent of its compressive strength, hence the need for tensile reinforcement.

The successful combination of steel and concrete is attributed to their very similar coefficients of expansion during normal temperature change, and the effectiveness of bond or adhesion between the two materials. Surface rust on steel will aid concrete adhesion, but loose flaky rust should be removed along with mill scale, mud and oil.

Chapter 13

Structural steelwork

Steel-framed construction has several advantages over traditional brick building and reinforced concrete. Offsite, factory prefabrication permits rapid and straight forward site assembly. It has less weight than any other material suitable for large buildings and multi-storey structures, hence the application of steel frames to poorer quality subsoils where a reinforced concrete frame would be too heavy. Large unobstructed spans permit design flexibility, and savings in foundation costs are another advantage of the light structural·weight. Construction time is shorter as steel has immediate strength, eliminating any delay waiting for the concrete to set and harden.

But steel-framed building also has its disadvantages, and these include the inability of the material to withstand the effects of fire. At temperatures of little over 500 °C a loaded section will lose strength sufficient to cause structural collapse. Exposed sections will rust and subsequently corrode without cladding or other suitable surface treatment, incurring additional costs; it may require periodic maintenance.

The advantages in many situations justify the disadvantages, and construction in accordance with the Building Regulations is deemed satisfied if the requirements of BS 449 : Part 2 : 1969, *The use of structural steel in building*, are applied. Hot-rolled standard sections are specified in BS 4 : Part 1 : 1980; these include the most common structural components, shown in Fig. 13.1: Universal Beams (UB), Universal Columns (UC), and Joists with tapered flanges, otherwise known as Rolled Steel Joists (RSJ).

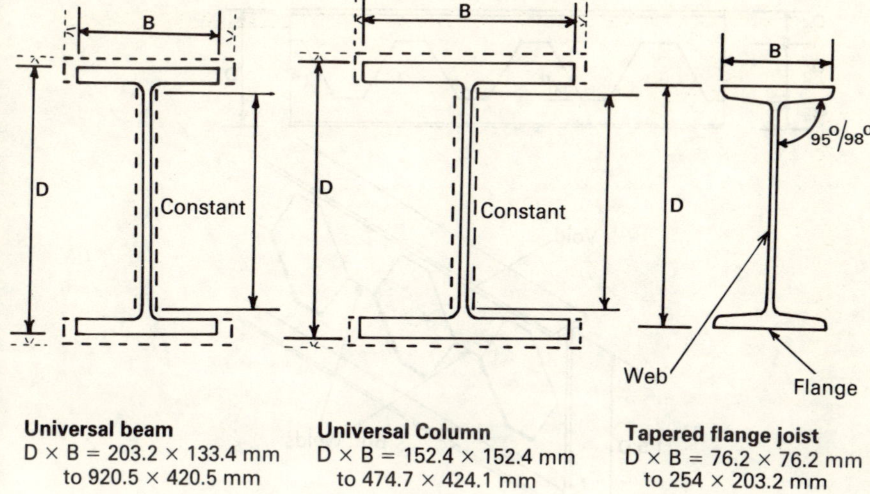

Fig. 13.1   Standard structural steel sections

Beam sections

These are mainly universal beams of parallel flanges, but tapered
flange joists exist to fulfil requirements for smaller sized beam sections.
Universal beams are rolled to a standard serial size, with variations in
flange and web thickness to provide several sections of different weight
for each serial size. Tapered flange joists have a small range of sizes,
and with the exception of only three serial sizes, have only one mass to
each serial size.

Castellated beam

Castellated beams are used to gain depth and resistance to deflection
with no increase in weight over a standard beam. A castellated line is
cut along the web of a standard section, separating it into two parts.
These are re-positioned as shown in Fig. 13.2 and welded to create a
new section 50 per cent deeper, having voids suitable for location of
ducting and other services beneath the floor or roof surface. Strength
of flanges is unimpaired and the resulting lightweight beam is still
capable of large spans if loading is not excessive.

Column sections

Universal columns fulfil all requirements for column sections, although
many old steel-framed buildings may be found with columns having
plates welded or riveted to the flange of a joist. These are known as
compound sections and are still occasionally used with columns and
beams where an exceptionally thick flange is required. Universal
columns are produced with standard serial sizes and specified with

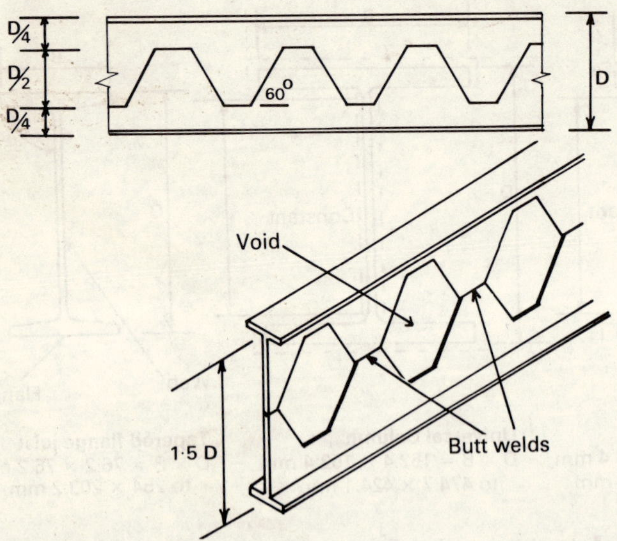

Fig. 13.2   Castellated beam

manufactured variables in flange and web thickness. These sections are easily identified from universal beams, as the majority have a flange width dimension very similar to the section depth, whereas universal beams and joists appear 'taller'.

Connections

Rivets were popular for both prefabricated connections and site use but have been virtually superseded by welding for workshop connections and bolts for site assembly. The process of riveting is uneconomic, particularly with the amount of skilled labour required to make the connection.

Welding

Welded connections are an electric arc process using an expendable steel electrode to fuse the connecting parts together. It is a relatively simple and quick procedure, eliminating the need for preliminary drilling of holes as rquired for rivets and bolts. Edge preparation for butt welds will be necessary for welding of thicker components, and an example is shown in Fig. 13.3. A fillet weld is also illustrated; these are often more convenient but less satisfactory for direct transfer of loads. Site welding, although unusual, may be employed, as it is more economical and simpler to use bolted connections on site.

Bolts

Three different types of bolt could be used for site assembly of

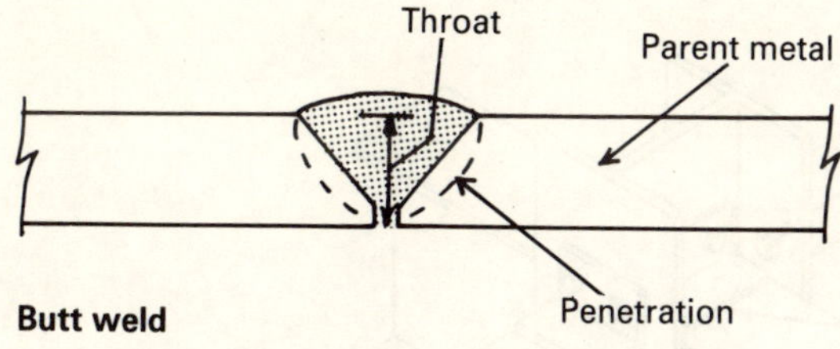

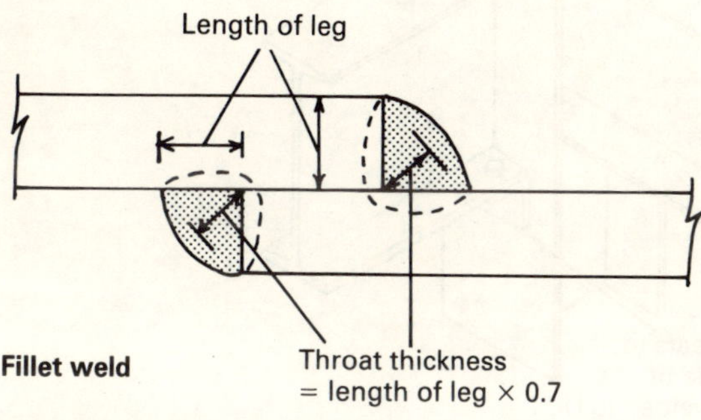

Fig. 13.3 Types of weld

steelwork; these are black bolts, machined bolts (turned and fitted) or high-strength friction grip bolts (torque bolts).

Black bolts

These are the cheapest to produce and provide the least precise fit, but are ideal for locating a section on a support bracket which provides direct bearing and resistance against shear. Manufacture is by forging and threads are the only part of the nut and bolt to be machined. Clearance between bolt shank and holes is approximately 2 mm.

Machined bolts

Machined or turned and fitted bolts are machined under the bolt head and turned along the shank. They provide greater accuracy of fit than black bolts with clearance of up to only 0.5 mm between bolt and hole. Dimensional accuracy of the steel frame is also greatly improved.

High strength friction grip bolts

These are manufactured from a higher yield steel than the other bolts and are also known as torque bolts because they are tightened in place to a predetermined extent by torque wrench. The amount of torque (Nm) is designed to transfer the load between connecting parts by friction. They are more expensive but have the advantage of transmitting greater load than other bolts, and the number and size of bolts required for each joint is less.

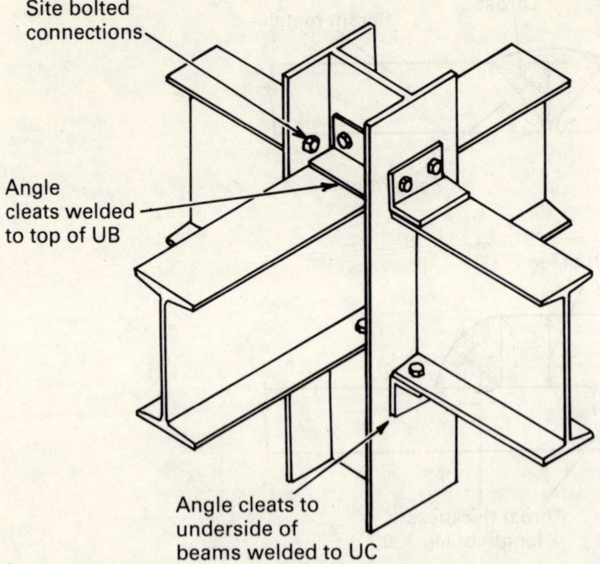

Fig. 13.4   Connection of beams to column

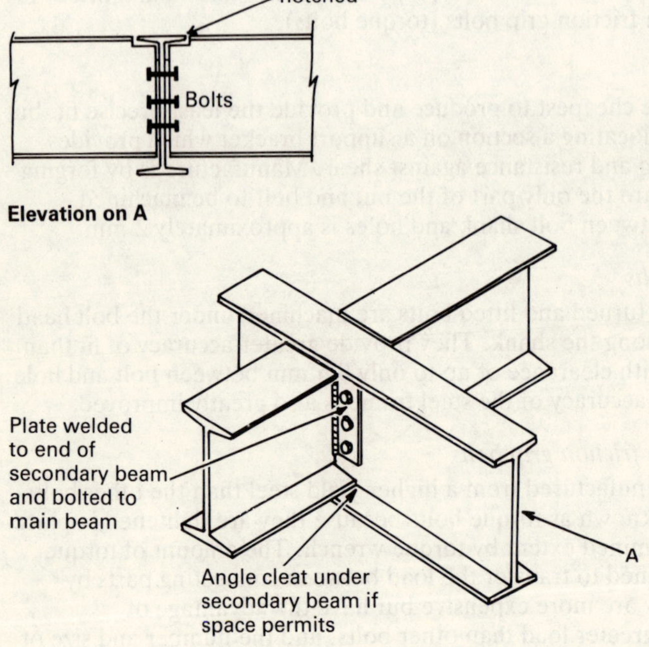

Fig. 13.5   Connection of secondary beams to main beam

Jointing and assembly

The method of jointing and assembling structural steelwork seldom
varies, and as a guide some simple examples are shown in Fig. 13.4
and 13.5. The load carried by beams should transmit to columns by an
angle or seating cleat welded to the column. Black bolts are provided
for location of members and a top angle cleat is provided for stability.
Here the angle is welded to the upper flange of the beam for
convenience of locating sections. Where secondary beams connect to
main beams a seating bracket should be provided if space permits.
Alternatively, web plates are welded to the secondary beam and bolted
to the main beam.

Foundation to columns

Columns or stanchions are provided with a steel plate welded to their
base as shown in Fig. 13.6. This is provided with holes for locating
projecting bolts from a concrete pad foundation. To ensure accurate
location of steel column on foundation, the bolts are positioned by a

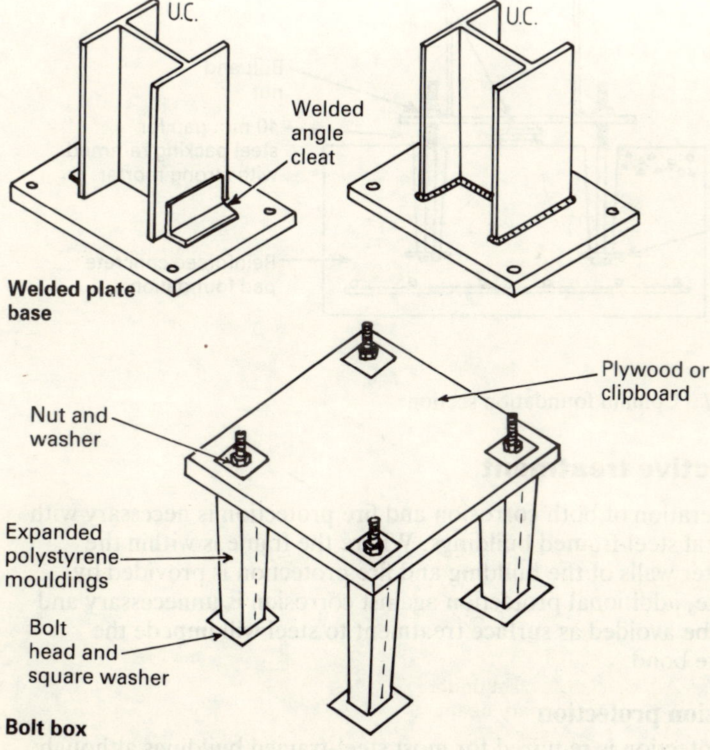

Fig. 13.6 Column base

'bolt box' within the wet concrete. This simple devise is also illustrated in Fig. 13.6. It consists of a sheet of plywood or chipboard drilled in the appropriate positions to match the holes in the column base, and bolts positioned with expanded polystyrene sleeves which dissolve with application of petrol, or are extracted when the surrounding concrete has set. Slight bolt movement is then possible to allow alignment with the column base. The column is aligned by providing steel levelling plates under the centre of the column base, and adjustment is by turning the nuts on the anchor bolts. Foundation detail and accessories are illustrated in Fig. 13.7; the space shown between base plate and foundation concrete is packed with a fairly dry cement and sand mortar in the ratio of 1 : 1.

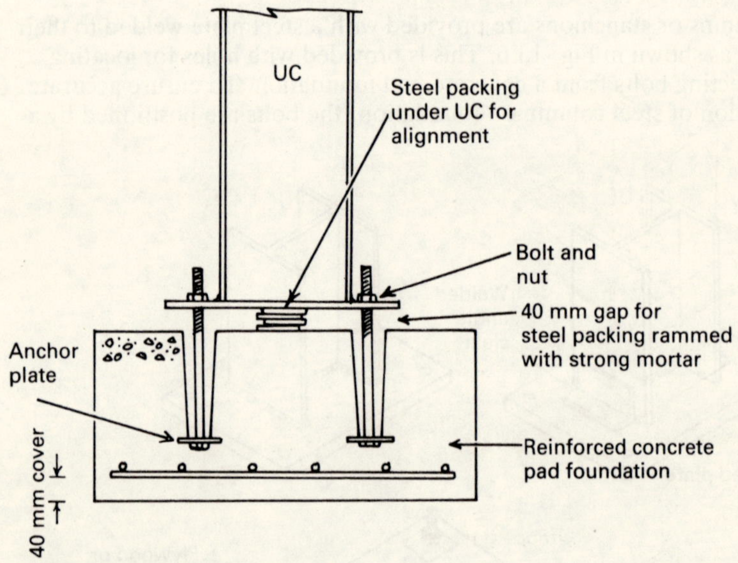

Fig. 13.7   Column foundation section

Protective treatment

Consideration of both corrosion and fire protection is necessary with structural steel-framed buildings. Where the frame is within the perimeter walls of the building and fire protection is provided by concrete, additional protection against corrosion is unnecessary and should be avoided as surface treatment to steel will impede the concrete bond.

Corrosion protection

Fire protection is required for most steel-framed buildings although some exemption is permitted in Part E to the Building Regulations.

This is generally single-storey buildings, and here corrosion protection will be of most significance. The choice of protection includes galvanising, metal spraying and painting. Galvanising is the most costly and effective, involving electro-plating with zinc after loose scale and rust are removed in an acid bath. Shot blasting is an alternative cleaning process and this may be followed by spraying with zinc or aluminium or by painting with red lead or zinc chromate primer. For additional protection to plated or sprayed sections, or for improved appearance, painting should be preceded by etch priming to seal the metallic surface.

Fire protection

As temperatures exceeding 500 °C can be achieved very quickly in a fire, the structural steel frame will require a protective casing.

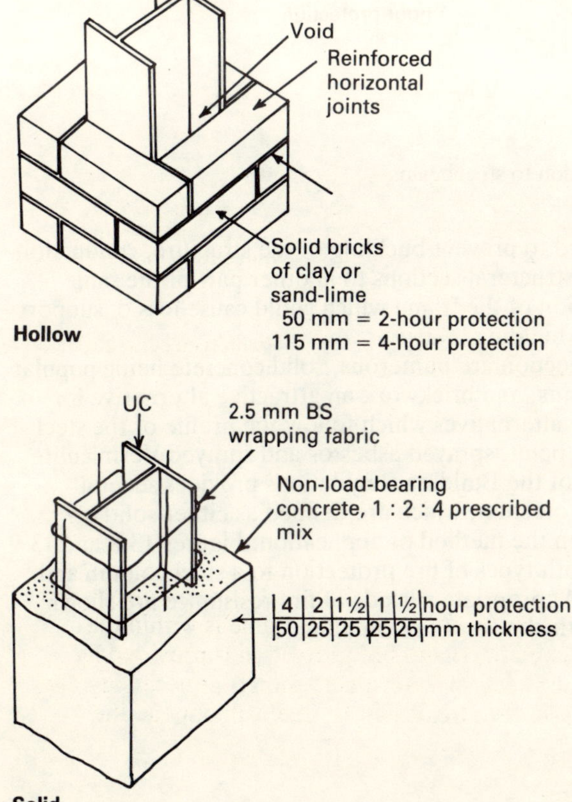

Fig. 13.8 Fire protection to steel stanchions

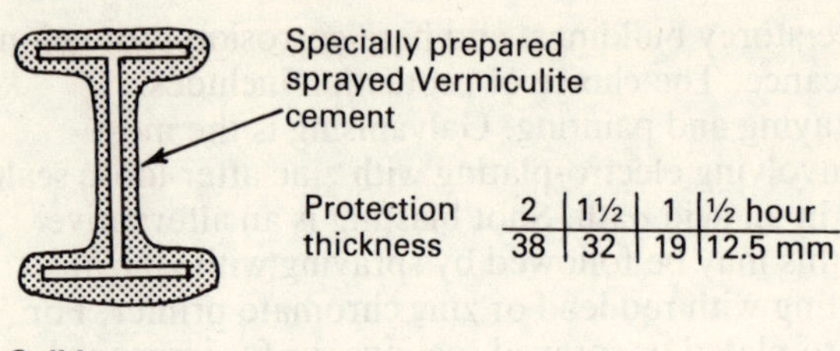

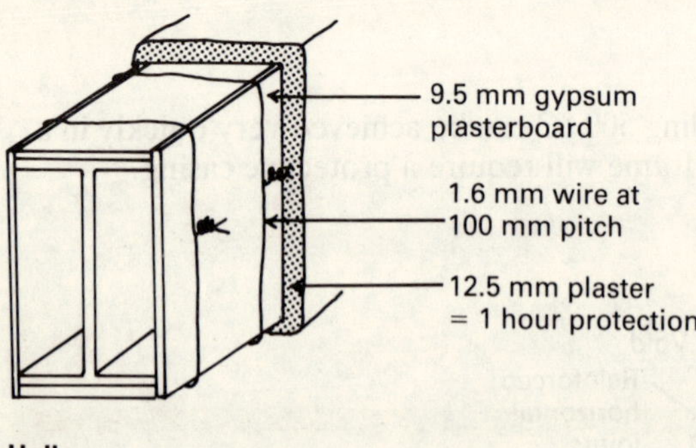

Fig. 13.9 Fire protection to steel beams

Protection is required to prevent buckling of the structure, conduction of heat through the structural sections to another part of the same building and expansion of the frame which could cause loss of support to other parts of the building.

Methods of protection are numerous, solid concrete being popular for beams and columns, and brickwork an attractive alternative for columns. Less bulky alternatives which follow the profile of the steel include intumescent paint, sprayed asbestos and sprayed Vermiculite cement. Schedule 8 of the Building Regulations provides several different forms of protection, which are defined as either solid or hollow, depending on the method of application. Figures 13.8 and 13.9 show examples of both types of fire protection to a steel column and beam respectively. Appropriate periods of fire resistance are shown with corresponding thickness of fire cladding.

Chapter 14

Natural and applied finishes

Wall and ceiling finishes

Plaster

Plaster is manufactured from gypsum rock, deposited in sandstone subsoils located mainly in Staffordshire, Derbyshire, Nottinghamshire, North Lancashire and parts of Sussex. Gypsum is the dihydrate of calcium sulphate, $CaSO_4.2H_2O$. It is crushed and then heated to remove 75 per cent of the water, to produce calcium sulphate hemihydrate, $2CaSO_4.H_2O$ or $CaSO_4.\frac{1}{2}H_2O$ to correspond with the term hemihydrate. Water is added on site, and after application during the final stage of setting the water returns to its original amount, $CaSO_4.2H_2O$, i.e. gypsum. Colour variations such as pink or grey result from impurities, but these do not affect the quality.

British Standard 1191 : Parts 1 and 2 : 1973 provide four classifications of gypsum building plasters: –

Class A. Plaster of paris.   BS 1191 Part 1. A hemihydrate which sets quickly, used mainly for crack filling and minor repairs.

Class B. Retarded hemihydrate.   Produced from a coarser gypsum than plaster of paris and has a retarder added by the manufacturer to slow the setting time.

British Standard 1191 : Part 1 type (a) undercoat plaster for mixing with sand.

1. Browning plaster.

2. Metal lathing plaster.
    British Standard 1191 : Part 1, type (b) finishing plaster.
1. Finish plaster
2. Board finish plaster
    British Standard 1191 : Part 2, premixed lightweight plaster
containing perlite or vermiculite, type (a) undercoat plaster.
1. Browning plaster
2. Metal lathing plaster
3. Bonding plaster
    British Standard 1191 : Part 2, type (b) finishing plaster.
1. Finish plaster

Class C. Anhydrous gypsum plaster.   British Standard 1191 : Part 1.
Produced by the dehydration of gypsum, $CaSO_4$. These plasters set
very slowly and harden some time after the initial set. They have poor
bonding qualities to anything but silicious or gypsum backgrounds, and
should therefore only be used as a finish.

Class D. Keene's plaster   British Standard 1191 : Part 1. An
anhydrous plaster of greater purity and hardness than Class C, being
more easily brought to a smooth, clean finish.

Background to plastered walls
The background is the surface to receive the first or undercoat of
plaster. The type of undercoat used will be determined by the suction
effect of the background, and these are classified according to high or
low suction effect. All backgrounds must be free of dirt and grease, and
where soot or carbon deposits have stained existing walls to be
replastered, these areas should be sealed with mastic paint sprinkled
with sand. Brick and block walls should have joints raked out to
provide a key, and the surface should be lightly dampened.
   Low suction backgrounds of dense concrete or engineering brick
should receive an undercoat of lightweight aggregate bonding plaster
or pre-treatment with a bonding agent. PVA proprietary bonding
agents or cement-based solutions are applied by brush to the clean
surface. Other difficult surfaces, including steelwork and painted walls,
may be plastered with the aid of galvanised wire bound to structural
steel and galvanised wire mesh secured to wall or ceilings with
galvanised nails at 100 mm intervals. Expanded metal lathing is an
alternative to mesh, and this is available as plain expanded or ribbed
expanded. It is less solid than other backgrounds, and must be
undercoated with Class B lathing grade plaster of 8 to 10 mm thickness
mixed with sand in the ratio of 1 : 1½/2. Finishing is normally with
2 mm of Class B finish grade plaster or Class C plaster.
   Cork, fibre board and expanded polystyrene should be surfaced
with tightly stretched galvanised chicken wire or expanded metal
lathing stapled at 100 to 150 mm spacing. Backing coat is with Class B

lightweight bonding plaster followed with a similar grade of finishing plaster.

Plasterboard

Plasterboard is a low-cost ceiling and wall lining material which may be finished with a Class B gypsum board finish plaster or decorated direct. The surface for decorating is ivory in colour and the other surface is grey with a double thickness of paper designed to receive gypsum plaster. Plasterboard is produced with a heavy paper finish and a solid or cellular gypsum core to the requirements of BS 1230 : Part 2 : 1970. Thickness is 9.5, 12.7 and 19 mm; the thinner board is suitable for timber stud walling and ceilings in domestic dwellings. Several standard sheet sizes are produced, the most popular being 2.4 × 1.2 m for 9.5 and 12.7 mm thickness. 9.5 mm thick laths with rounded longitudinal edges 1.2 m long and 0.406 m wide are also widely used where access and location is difficult with larger sheets. The edges of boards are paper bound as shown in Fig. 14.1 and ends are cut, exposing the plaster core.

Plasterboard is secured with 30 mm galvanised steel clout nails at 150 mm spacing, and where 9.5 mm board is used, sagging is prevented by spacing the joists at a maximum of 400 mm. Boards are cut at the centre of the joist, leaving a 5 mm gap between adjacent boards. Joints are made good with board finish plaster worked into the

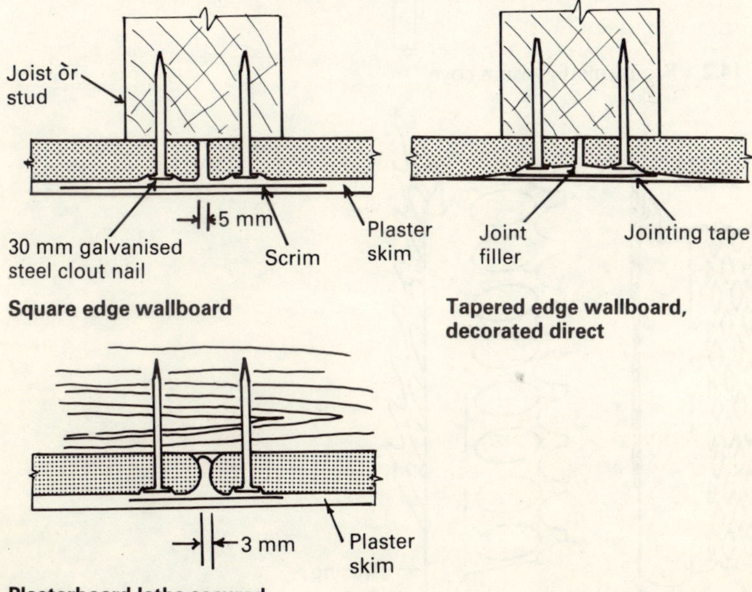

Fig. 14.1   Plasterboards

gap and spread about 50 mm either side of the joint. The joint is bridged and reinforced with canvas or jute scrim trowelled in.

Internal angles should be reinforced with scrim to avoid shrinkage cracks. An attractive alternative is a cornice or manufactured gypsum cove nailed or glued to the ceiling and wall as shown in Fig. 14.2. External angles may be reinforced with metal angle bead shown in Fig. 14.3. These are galvanised expanded metal with projecting nosing plumbed vertical and secured with dabs of plaster at 500 mm intervals. Undercoat or floating coat is finished just below the nosing with the finishing coat brought up level.

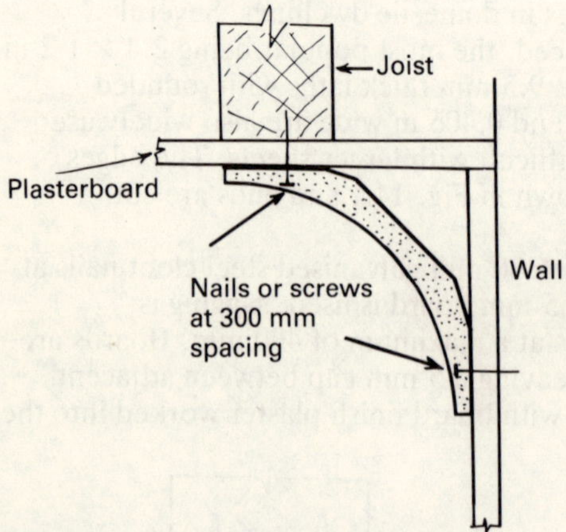

Fig. 14.2   Pre-formed gypsum cove

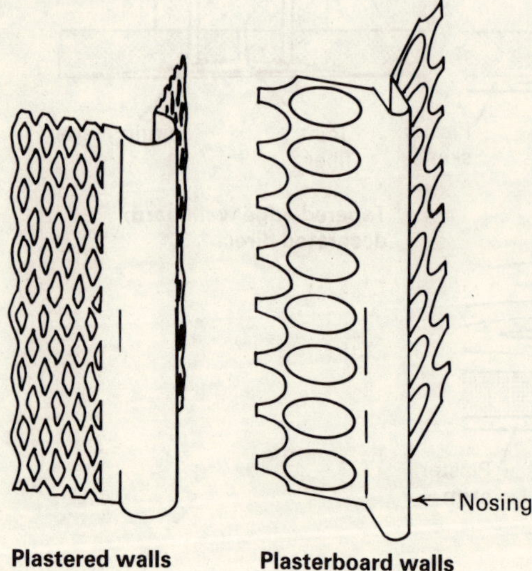

Fig. 14.3   External angle bead

Application

The plasterer's tools are numerous and many craftsmen have implements tailored to their own needs. A few basic tools are illustrated in Fig. 14.4. Plaster is mixed to the correct consistency and placed on a spotboard stood close to the work area. Small quantities are transferred by trowel to the hawk and smaller quantities are taken off the hawk and trowelled to the background. The second coat is applied by skimming float and final coat by trowel. Where only two-coat work is required, the first coat is floated and final coat trowelled. Rendered undercoats are always keyed to support further coats, and a popular method is by devil float, manufactured by projecting nails through the surface of an old skimming float. Irregular applications are levelled with a timber straight edge or darby and internal and external angles finished with purpose-made steel trowels.

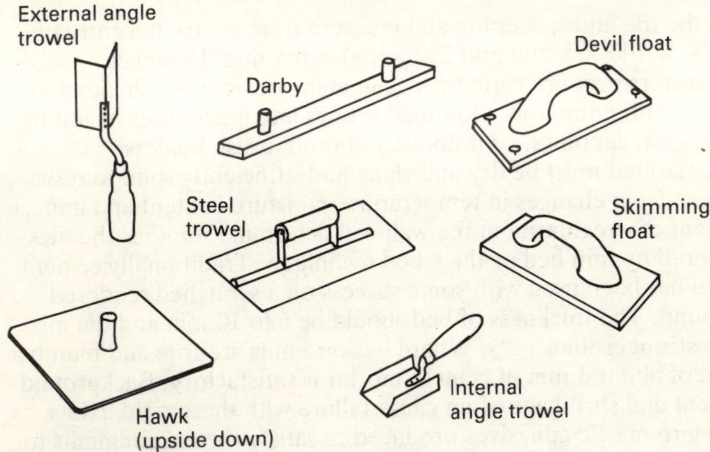

Fig. 14.4 Plasterer's tools

Glazed ceramic tiles

Glazed ceramic tiles are a kiln-fired earthenware product glazed on one surface and exposed edges (see Fig. 14.5). A wide range of colours are available with variations in pattern and surface profile. They are resistant to staining and chemical attack and provide an ideal wall finish to kitchens and bathrooms.

British Standard 1281 : 1974 provides details of non-modular and modular preferred sizes. Non-modular sizes are based on imperial dimensions and metricated, the nominal sizes are $152 \times 152 \times 5$, 5.5, 6 and 8 mm thick, and $108 \times 108 \times 4$ and 6.5 mm thick with spacer lugs. A 152 mm square tile is also available 9.5 mm thick without lugs.

Modular tiles to the British Standard preferred basic size are produced to co-ordinate with metric dimensions. These dimensions

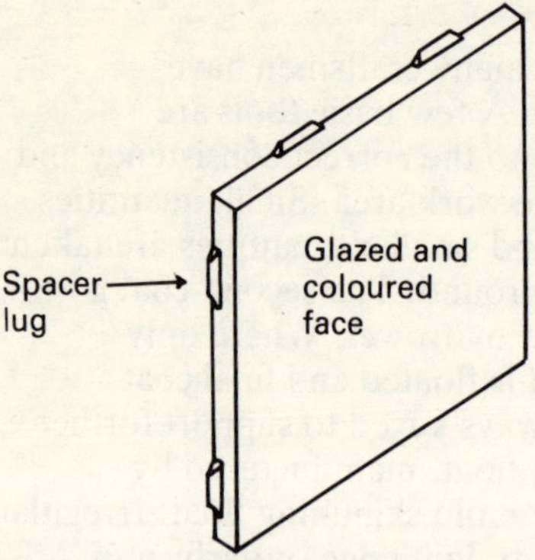

Fig. 14.5   Glazed ceramic tile

include the tile and spacer lug and measure from centre to centre of joint: 100 × 100 × 5 mm and 200 × 100 × 6.5 mm. However, manufacturers produce ceramic tiles in many other sizes, shapes and thickness, with numerous additional fittings and accessories including coves, angles, cappings, coat hooks and tooth brush holders.

Background must be dry and clean and sufficiently solid to resist movement from changes in temperature, moisture content and impact. The extent of irregularity in the wall will determine whether the tiles are adhered by thin bed or thick bed technique. Traditionally cement and sand has been used with some success on a scratched rendered background. The thickness of bed should be 6 to 10 mm and the mix ratio no stronger than 1 : 3. Where backgrounds are true and plumb a thin bed of about 3 mm of cement mortar is satisfactory. Background movement and shrinkage often cause failure with these rigid fixings and modern mastic adhesives produced to satisfy the requirements for wall tiling specified in BS 5385 : Part 2 : 1978 are often more suitable. Many adhesives are suitable for difficult surfaces such as chipboard and existing tiled surfaces, and variations are available which will accommodate surface irregularities up to about 12 mm.

Joints are grouted or pointed at least 24 hours after tile fixing. Grout is a proprietary white cement worked into the joint with finger or soft cloth. Wide joints are filled and pointed with a small trowel. Where sterility is very important, epoxide-based grouts are preferred to provide a hard impervious finish.

Paint

Painting is required for decoration, preservation, protection and hygiene. Paint is basically a pigmented liquid which dries and hardens on application and many types are manufactured for practically every surface. A protective barrier is built up from primer, undercoat and finish.

The primer is the first coat and must be carefully selected for good adhesion to the particular background. Some specifically even out the suction effect on porous surfaces and others control rusted surfaces and seal metals for further treatment. The undercoat reinforces and builds up the paint thickness by good adhesion to the primer. It also levels and obliterates minor surface imperfections. The finishing coat provides protection, colour and texture or surface smoothness. The extent of smoothness is defined in BS 4800 : 1981 as full gloss, semi gloss, eggshell or matt. The British Standard provides 100 standard colours including black and white which are derived from the Munsell scale of colours. This is an international system providing accurate description of hundreds of colours, based on three properties, hue, value and chroma. Hue refers to the basic colour, value measures lightness or darkness and chroma compares intensity of hue with greyness.

Paint composition

The components of paint vary considerably to suit different surfaces, conditions and finish. All paints contain most if not all the ingredients shown in Fig. 14.6. The binder, also known as the vehicle, is an oil or varnish which binds the other ingredients together and holds the particles of pigment in suspension. It is the most important part of the paint and must dry and harden at a rate suited to method of application.

Pigments are the solid components of paint and are obtained from a considerable variety of sources. Many are metallic; white lead and iron oxides are two common examples and others are coloured earths, e.g. yellow ochre, or chemical compounds, e.g. prussian blue. Extenders are also a solid component of finely ground minerals such as china clay. They are generally white in colour but transparent in oil

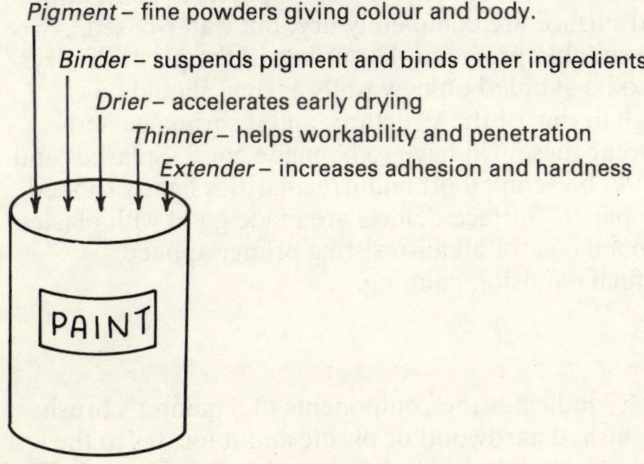

Fig. 14.6 Components of paint

and have the effect of easing application, increasing hardness and adhesion.

Drying agents and catalysts evaporate, absorb oxygen and convert to a solid by oxidisation. They are soluble compounds of metals, e.g. lead and manganese in linseed oil and white spirit, having the effect of accelerating early drying. Solvents are included in the paint manufacturing process, but additional thinning may be required to suit site conditions and method of application. Thinners are colourless and selected and applied in accordance with manufacturer's recommendations. They are a volatile liquid which evaporates quickly during the early drying process. Most oil paints may be thinned with white spirit and emulsions with water.

Background

Surfaces most frequently painted in housing are wood and plaster. Treatment of woodwork and joinery varies with the type and nature of timber, and the following procedure is to be considered the minimum for adequate protection and decoration. Initially wood must be dry, clean and free from irregularities which light sanding should remove. Knots contain resin and will discolour paint if not sealed with knotting compound, a mixture of shellac and methylated spirit. Priming is traditionally with a pink primer, basically a blend of red and white lead, but acrylic polymer emulsion primers are often preferred because of their non-toxic nature. Aluminium-based primers are also popular, automatically sealing knots. Defects and cracks are filled with a patent stopping compound or putty and two coats of undercoat should follow when stopping is hard. Finishing may be one coat only for interior use, but at least two coats should be applied externally. For quality finishes, each coat of paint must be lightly sanded before application of the next coat.

New plaster should not be treated with an impervious paint until the background and surface are completely dry, but water-based emulsion paints are suitable provided the surface of the plaster is dry. Papering should also be avoided on new walls as time should be allowed for thorough drying of the structure, and if shrinkage and settlement cracks occur these will have to be made good. Splashes and drips of plaster should be scraped off and irregularities lightly rubbed down with abrasive paper. Surface defects are made good with plaster filler and a well thinned coat of alkali-resisting primer applied throughout before final emulsion painting.

Application

Brushes.   Figure 14.7 indicates the components of a painter's brush. The handle is of varnished hardwood or plastic and it locates to the filling with a riveted or pressed nickel-plated steel ferrule or stock. The setting is an adhesive, usually epoxy resin, which secures the brush

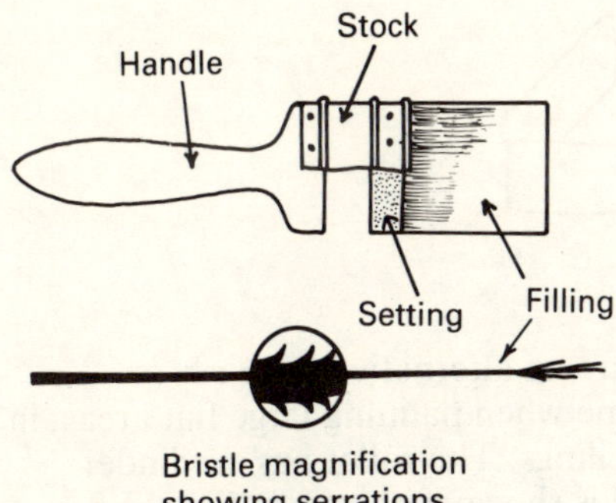

Fig. 14.7 Paint brush detail

filaments together in the stock and the filling is either natural hair or fibres or man-made fibres.

Pure bristle or hog hair is an excellent filling material having the ability to hold more paint than other materials. This is because its strength is sufficient to allow a long length, and minute serrations along each hair prevent the hairs lying too close together. Further properties include a slender taper throughout the hair length with split or flagged ends providing a soft tip for even distribution of paint.

Horse mane or tail is a cheaper filling but has no natural taper or split ends. It is soft and lacks the strength of bristle. Natural glass fibres are coarse and lack the properties of bristle; they are used mainly for washing down brushes or may be mixed with bristle or horse-hair. There are many synthetic fibres including nylon. These are good imitations of bristle, having taper and flagging but lacking serrations, they are hard wearing and a cheaper alternative as natural bristle becomes scarcer and more expensive. Several examples of popular brushes are shown in Fig. 14.8.

Fig. 14.8 Brushes

Fig. 14.9   Cylinder roller and tray

Cylinder roller.   Cylinder rollers are an alternative to brush
application, saving considerable time when painting large flat areas, in
particular emulsioning walls and ceilings. The roller has a cylinder
attached to an axle and handle and is shown with tray in Fig. 14.9.
Cylinders have replaceable covers produced from plastic foam,
synthetic filaments, mohair or lambswool.

Floor finishes

Floor finishing techniques are numerous and many, including
carpeting, are no normally part of the general contractor's work. The
following section includes a few of the more popular finishes that a
contractor would be expected to install. The functional requirements
governing the choice of finish will include cost, appearance, durability
and comfort. Amount of maintenance is another important
consideration and suitability and effectiveness are two other criteria
often disregarded.

Timber

Planed softwood boards have been the accepted surface for suspended
timber floors until the introduction of chipboard as a cheaper
alternative. Softwood boards were never intended as the floor finish,
but sanded and sealed with varnish they provide an effective, natural,
practical alternative to the expense of carpeting. They will withstand
moderate use and regular wax polishing will prolong the life of the
varnish. Secret nailing shown in Fig. 14.10 provides a better finish,
uninterrupted by nail heads, and Fig. 14.11 indicates methods of
securing boards to a concrete or screeded floor.

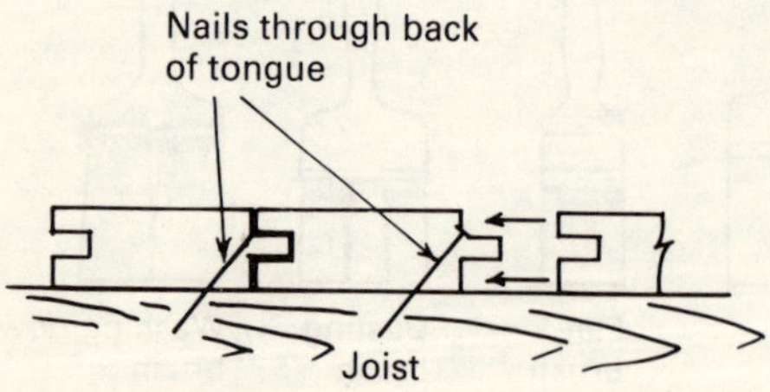

Fig. 14.10   Secret nailing to tongued and grooved boards

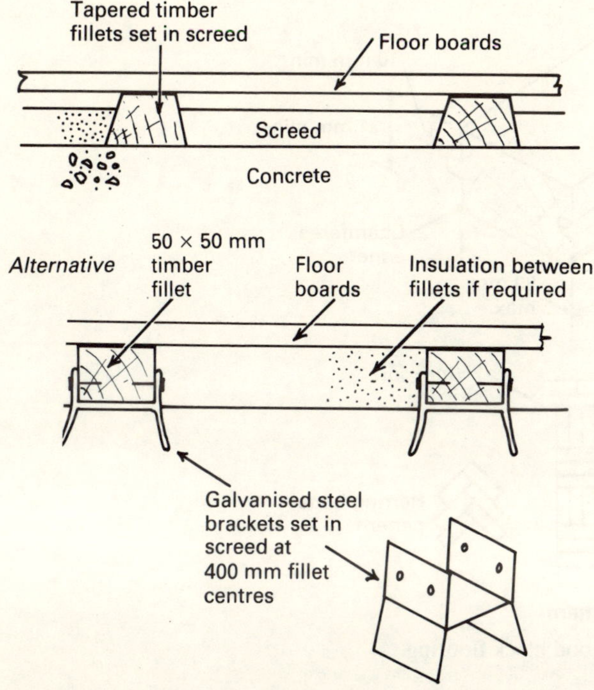

Fig. 14.11 Timber floor finish to concrete floors

Hardwood floors are provided in strip or block form and selected hardwoods are used to provide block parquet flooring. Hardwood strips are available in nominal widths up to 100 mm, with tongued and grooved edges and possibly matched ends. Fixing is by secret nailing. Hardwood blocks are manufactured to the requirements of BS 1187 : 1959 shown in Fig. 14.12. They are secured with hot or cold bituminous solution to a primed and wood floated screed. The arrangement of blocks may be herringbone or basketweave pattern with a cork expansion strip located close to the perimeter.

Parquet blocks are produced at least 6 mm thick and backed with foil or felt. They are mounted on plywood or other suitable sheet material to form panels, usually in basketweave pattern, which are glued and pinned to the subfloor.

Cork tiles

Cork tiles are available with a transparent vinyl finish to increase durability. Without this finish they will require sealing with polyurethane varnish. The subfloor must be clean, dry and level, and if imperfections exist an underlay of hardboard will be required. Thickness of tile is between 3 and 6 mm, and 300 × 300 mm is a

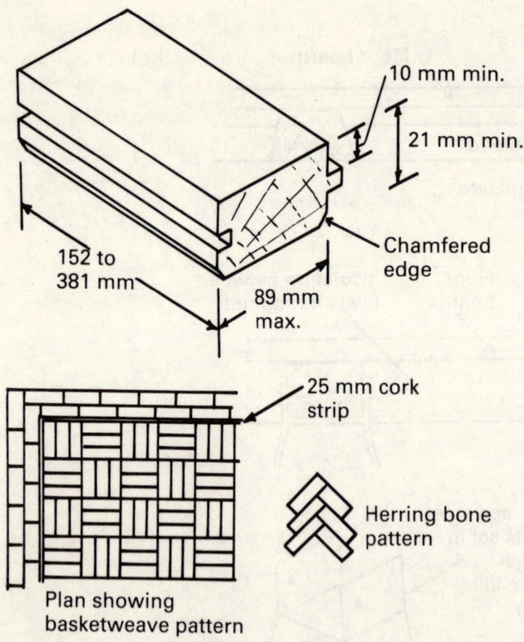

Fig. 14.12 Hardwood block flooring

common surface size. Carpet or sheet cork is also available and fixing
for both tile and sheet is by adhesive.

Plastic tiles

In modern form these originated in the late 1940s as a thermoplastic or
asphalt tile secured to solid subfloors with a bituminous adhesive.
These tiles are more brittle and less easy to lay than the flexible PVC
tiles which have superseded them. Colour range and quality of PVC
tiles is superior, and maintenance is easier. Impact sound transmission
may be reduced by backing with resilient material, but this is normally
only available with the sheet variation. PVC tiles are produced to BS
3261 : Part 1 : 1973 with thicknesses of 1.5, 2, 2.5 and 3 mm. Standard
sizes are 225, 250 and 300 mm square, and fixing is by manufacturer's
recommended adhesive. A cheaper variation with fillers of asbestos
fibre and limestone is available. These are known as vinyl asbestos tiles
produced to BS 3260 : 1973. They have good resistance to chemicals
and moisture, but are less flexible than PVC, more prone to marking
and less easily cleaned.

Clay or quarry tiles

Clay tiles are cold and noisy under impact but their dense nature
provides excellent wearing qualities. British Standard 1286 : 1974
includes both ceramic floor tiles, produced by compaction and

blending of ceramic powders and kiln firing, and clay floor quarries manufactured by kiln-firing extruded or moulded refined clays. They are produced in many sizes and thicknesses. The preferred dimensions are 200 × 100 mm and 100 × 100 mm with 9.5 mm thickness for ceramic tiles and 19 mm thickness for quarries. Fixing is by bedding in cement and sand mortar no richer than 1 : 4 and at least 20 mm deep. When covering large areas, the risk of differential movement between tiles and subfloor is reduced by using a separating layer of polythene or building paper as shown in Fig. 14.13. Here the bedding may be reduced to 12 mm and an expansion joint should be provided at the perimeter.

Tiled roof finishes

Roof tiles are manufactured from clay or concrete, in double lap or single lap patterns.

Plain tiling

Tiles laid with a double lap are otherwise known as plain tiles of standard size 265 mm long × 165 mm wide. An example is shown in

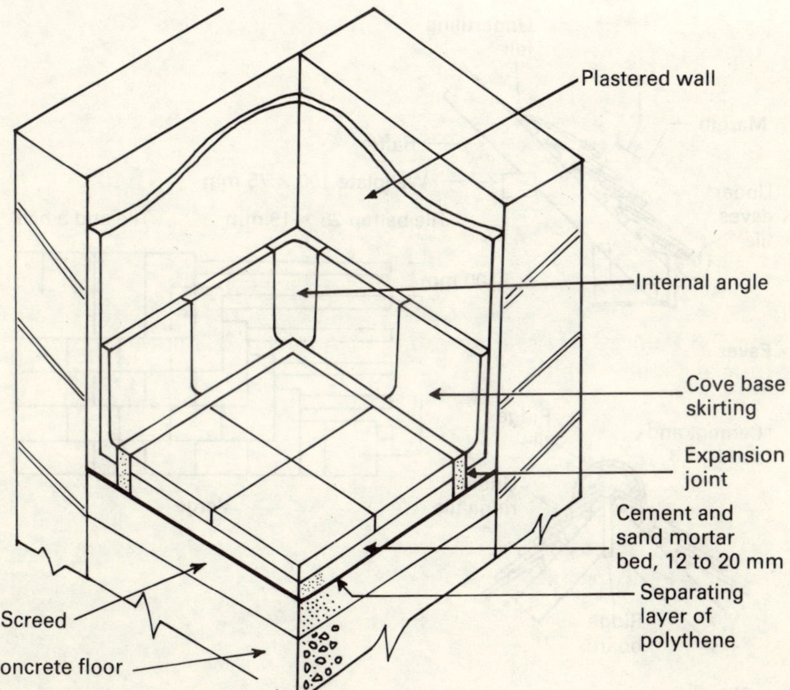

Fig. 14.13   Clay floor tiles

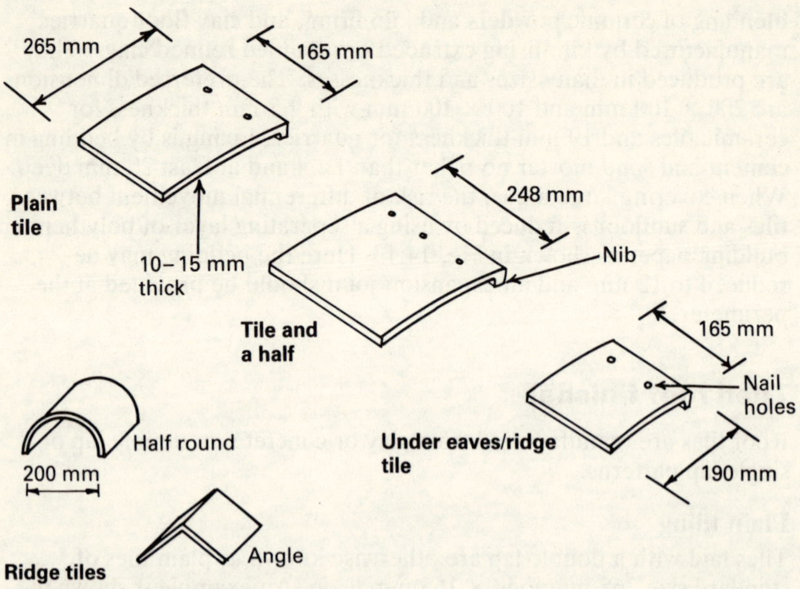

Fig. 14.14 Plain tiles

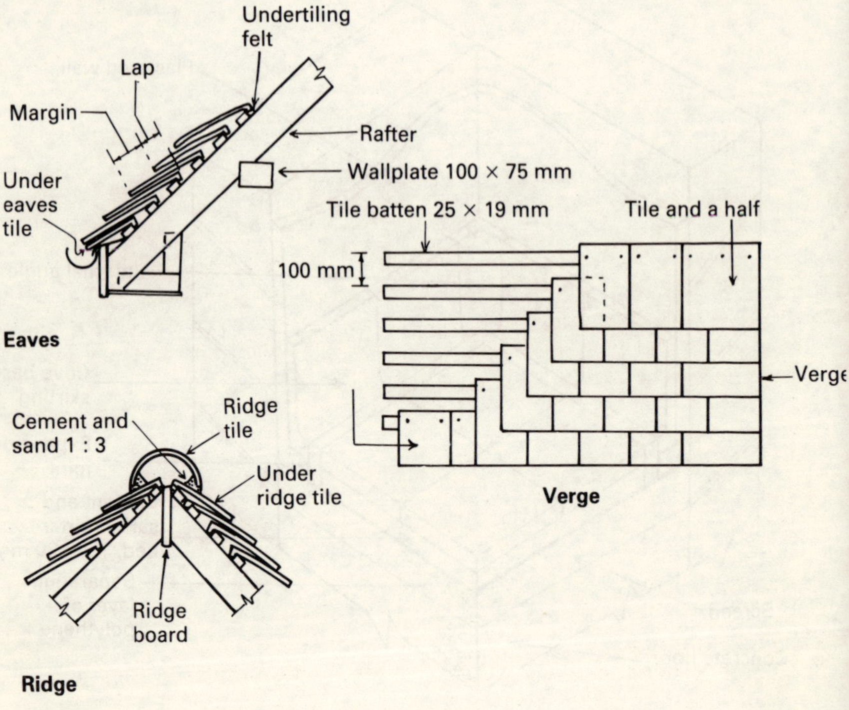

Fig. 14.15 Plain tiling

Fig. 14.14, with several variations. The tile and a half is used on alternate courses at gable ends and abutments, and the under eaves/under ridge tile below the first exposed course at gutter level, and below the ridge tile at the roof apex, respectively. Figure 14.15 illustrates the location and construction associated with these special plain tiles.

Plain tiles are cambered to prevent water penetration by capillarity and each tile is laid to overlap the joint between the two tiles below. For effective weather exclusion the roof structure should be pitched at no less than 40° to provide a tiling angle of at least 35° to the horizontal. Support is by 25 × 19 mm softwood battens where the rafter spacing does not exceed 450 mm, and 32 × 25 mm where rafters are spaced 450 to 600 mm. The spacing or gauge of the battens is normally 100 m, and this is derived from the formula:

$$\text{Gauge} \quad = \quad \frac{\text{length of tile} - \text{lap}}{2}$$

Lap should never be less than 65 mm, and this represents the amount each tile overlaps the tile but one below. In conditions of severe exposure a lap of 75 to 90 mm would be more suitable. Nailing of tiles is every fourth course, increasing to every third course in very exposed conditions and on roofs pitched up to 55°. Over this angle, every tile should be nailed. Galvanised steel nails are popular, but aluminium alloy nails offer greater resistance to corrosion. A layer of untearable waterproof felt is provided between the battens and rafters, laid parallel to the battens with overlaps of at least 150 mm. Its function is to prevent driven rain and snow penetrating the roof below and between tiles, to improve the thermal insulation and to provide a waterproof layer if tiles become damaged or broken.

Hips and valleys require special treatment, and two possibilities are shown to each situation in Fig. 14.16. Ridge tiles extending down the hip terminate with a galvanised hip iron, but a more attractive finish is provided with bonnet tiles. Valleys too have many variations, the simplest treatment is with purpose-made valley tiles or the more expensive lead lining.

Single lap tiling

The shape and size of single lap tiles is extensive, and in addition to the single head lap they have a common side lap. In modern tiles the side lap interlocks, providing a positive location to adjacent tiles, and head lap should never be less than 75 mm. Some types of single lap tile are suitable for roofs pitched as low as $17\frac{1}{2}°$, but because of their size, battens are 38 × 19 mm on rafters up to 450 mm spacing and 38 × 25 mm where rafter spacing is up to 600 mm. Each tile is secured with a clip or nail depending on design of tile, and examples of tiles with construction at ridge, eaves and verge are shown in Fig. 14.17 and 14.18.

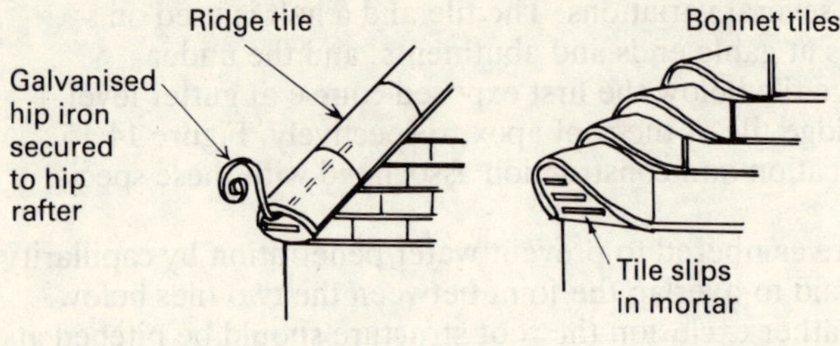

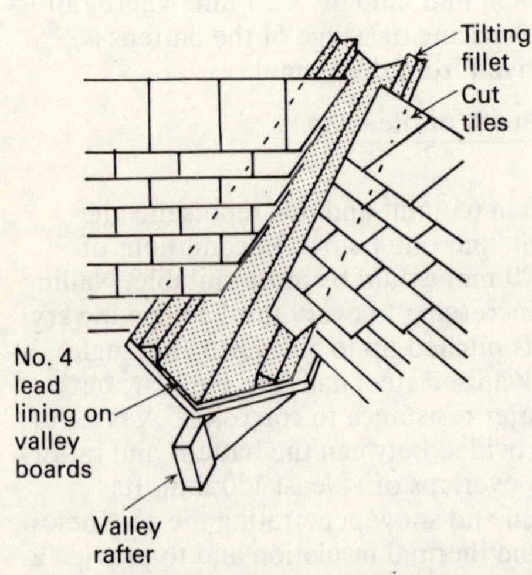

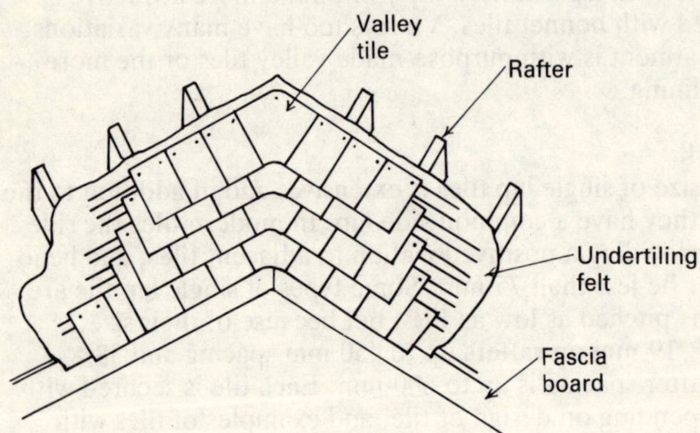

Fig. 14.16 Plain tiling at hips and valleys

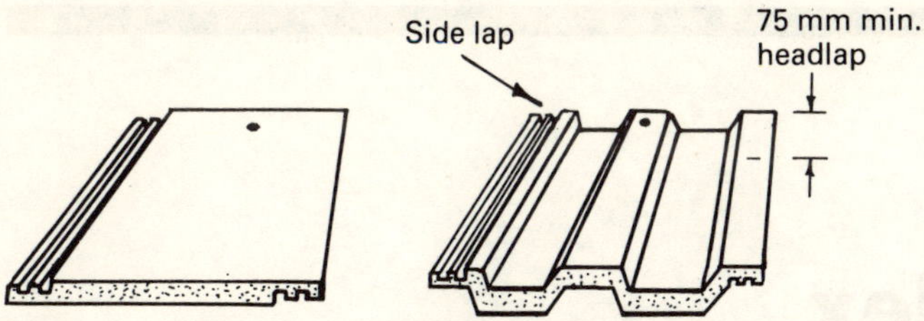

Fig. 14.17 Examples of single lap concrete tiles

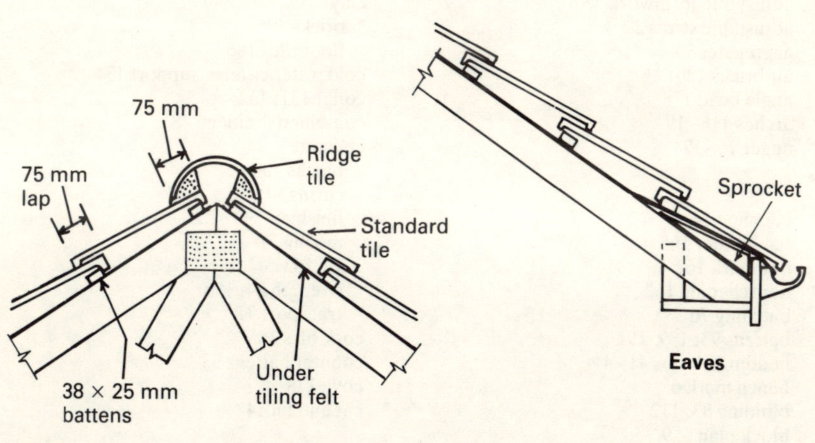

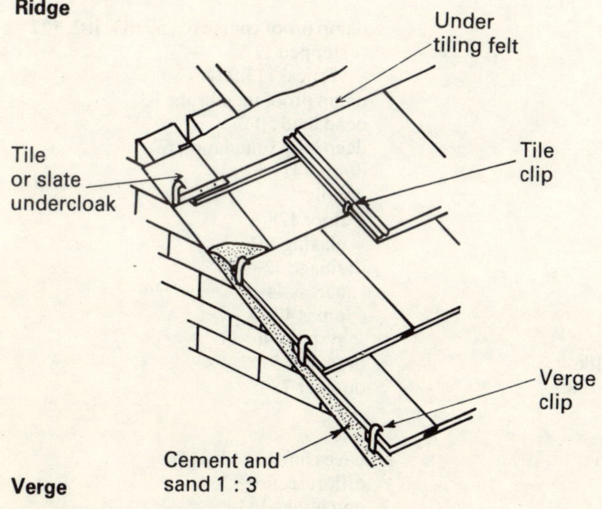

Fig. 14.18 Single lap tiling

Index